MODELING RUSSIA'S ECONOMY
IN TRANSITION

To my dear parents,
Dr. Heinz and Uta Wehrheim (neé Bilger),
whose support has been essential for my education and for creating my
curiosity and interest in political issues and in other countries.

Modeling Russia's Economy in Transition

PETER WEHRHEIM
Center of Development Research, University of Bonn

ASHGATE

Published by
Ashgate Publishing Limited
Gower House
Croft Road
Aldershot
Hants GU11 3HR
England

Ashgate Publishing Company
Suite 420
101 Cherry Street
Burlington, VT 05401-4405
USA

Ashgate website: http://www.ashgate.com

British Library Cataloguing in Publication Data
Wehrheim, Peter
 Modeling Russia's economy in transition. - (Transition and development)
 1.Russia (Federation) - Economic conditions - Mathematical models 2.Russia (Federation) - Economic conditions - 1991-
 I.Title
 338 .9'47'0015118

Library of Congress Cataloging-in-Publication Data
Wehrheim, Peter.
 Modeling Russia's economy in transition / Peter Wehrheim.
  p. cm. -- (Transition and development)
 Includes bibliographical references and indexes.
 ISBN 0-7546-3299-7
  1. Russia (Federation)--Economic conditions--1991--Mathematical models. I. Title. II. Series.

 HC340.12 .W44 2003
 330.947'001'1--dc21

                                                         2002034530

ISBN 0 7546 3299 7

Printed and bound in Great Britain by Antony Rowe Ltd., Chippenham, Wilts.

Contents

List of Figures

List of Tables

Acknowledgements

This book has been written while working at the Center of Development Research (ZEF) and the Institute for Agricultural Policy, Market Research, and Economic Sociology at the University of Bonn, Germany, between 1998 and 2001. The manuscript has been completed while I was affiliated with the Chair of Agricultural and Economic Policy at the Faculty of Agriculture, University of Bonn and with the Centre of Institutional Reforms and the Informal Sector (IRIS) at the University of Maryland, College Park, Maryland. The support I received from all three research institutes in implementing the research and writing this book is gratefully acknowledged. The idea and the beginning of the research endeavor to model Russia's economy in transition, however, date back longer: it was part of a research project conceptualized by Joachim von Braun in 1994 when he was Professor for Food Economics and World Food Economics at the University of Kiel, Germany. His intellectual support and his guidance as a mentor have been a *conditio sine qua non* for the successful completion of this research and book project. The financial support for this project was provided by the *VolkswagenStiftung*, Germany. This is gratefully acknowledged.

The task of modeling Russia's economy in transition was non-trivial and without the discussions with the 'CGE-experts' Sherman Robinson (IFPRI), Peter Wobst (IFPRI), and most of all with Manfred Wiebelt (Kiel Institute for World Economics) it would have been a daunting one. On top, I received a lot of support from 'inside' Russia for a better understanding what transition is about and in compiling the respective data-base. While acknowledging the skepticism of my Russian colleagues and friends about even trying to model the Russian economy in transition I would like to express my gratitude to those who encouraged me to continue my work in spite of their doubts: special thank goes to Eugenia Serova (Institute for the Economy in Transition) who has always been ready for open discussions and who thereby has helped me in shaping this study. I would also like to thank Irina Khramova, Olga Melyukhina (Institute for the Economy in Transition and OECD, respectively), Sergei Kiselev (Lomonosov State University, Moscow) and colleagues at the Agrarian Institute in Moscow for valuable academic and many inspiring non-academic discussions. Furthermore, I would like to thank colleagues who have commented on earlier drafts of this book and the research presented therein: Klaus Frohberg, Tom Hertel, Dieter Kirschke, Ulrich Koester, Arnim Kuhn, Wilhelm Henrichsmeyer, and Susanna Wolf. Furthermore, I owe thanks to many research colleagues at various institutions, near and far, who have discussed with me earlier versions of the model which I have presented at various conferences and seminars.

I am also grateful to those at the Center of Development Research who not only commented 'academic-wise' but also gave 'moral' support: representative for all junior and senior researchers, I would like to mention my colleagues Johannes Jütting, Ulrike Grote, and Daniela Lohlein. By the same token I have received great support from my 'big' and my 'small' family who always believed in me and if not did not show it. While it is impossible to list the names of all members of my 'big family' I am indebted to my fiancée who today is my wife, Simone Lünenbürger, and, to our son, Moritz Karl Wehrheim, for sharing my time with this book project while I was writing and polishing the manuscript.

List of Acronyms

ACGE	Applied Computable General Equilibrium
AMS	Aggregate Measure of Support
bR	Billion rubles
CEEC	Central and Eastern European Countries
CES	Constant Elasticity of Substitution
CET	Constant Elasticity of Transformation
CGE	Computable General Equilibrium
CIS	Commonwealth of Independent States
c.p.	*ceteris paribus*
CPI	Consumer Price Index
ECCI	Euro-money Country Creditworthiness Index
EU	European Union
FAO	Food and Agricultural Organization of the United Nations
FSU	Former Soviet Union
GAMS	General Algebraic Modeling System
GAO	Gross Agricultural Output
GATT	General Agreement on Tariffs and Trade
GDP	Gross Domestic Product
GE	General Equilibrium
GNP	Gross National Product
GOSKOMSTAT	Statistical Office of the Russian Federation
GTAP	Global Trade Analysis Project
IET	Institute for the Economy in Transition
IFPRI	International Food Policy Research Institute
IfW	Institute for World Economics
IMF	International Monetary Fund
IOC	Input-Output-Coefficients
IOT	Input-Output-Table
LAE	Large Agricultural Enterprises (Former collective farms)
LPH	*Lichnie Podsobnie Khozyaistva* (Private subsidiary plots)
NMP	Net Material Product
ODP	Origin of Domestic Production
OECD	Organization for Economic Cooperation and Development
PSE	Producer Subsidy Equivalents
Rb	Ruble
RF	Russian Federation
SAM	Social Accounting Matrix
SNA	System of National Accounts
t	(metric) tons

TFP	Total Factor Productivity
TOT	Terms of Trade
tR	Trillion rubles
UDP	Use of Domestic Production
WTO	World Trade Organization
ZEF	Center for Development Research

Chapter 1

Introduction

The *transition process in Russia*, which started vigorously at the beginning of the 1990s, has yielded *negative economic growth rates* for almost a decade, with the minor exceptions of 1997 and 1999. Shortly before and after the world financial crisis which hit Russia late in 1997 and culminated after mid-August 1998, total Gross Domestic Product (GDP) grew moderately (by 0.5% in 1997 and 1.5% in 1999; WORLD BANK 1999 and 2000). In the mid–1990s and up to 1997, macroeconomic conditions stabilized in Russia (Table 1-1): inflation was reduced; the government budget deficit reached its highest level in 1994 and then decreased; and the exchange rate became stabilized. At the same time, some macroeconomic indicators such as the official unemployment rate deteriorated further as the former state owned enterprises released an increasing amount of labor. In contrast to many other transition economies, the overall trade balance of the Russian Federation (RF) remained positive during the 1990s because of substantial raw material exports, mostly consisting of gas and oil.

Two related sectors were particularly suffering from the ongoing transition process and the associated restructuring of the economy: *agriculture and the domestic food industry* (see Table 1-2). In fact, agriculture's performance also deteriorated continuously until 1997 and was affected negatively in 1998 by the financial crisis as well. In particular, the formal rural credit system collapsed almost completely in the course of the crisis. Very moderate signs of recovery in agriculture were notable in 1999, when the real devaluation of the ruble against major western currencies contributed to an increased competitiveness of domestic food producers in relation to foreign competitors.

During the first transition decade not only the *share of agriculture in GDP declined* from more than 16% in 1990 to about 6.5% in 1998 but also agriculture's share in total national investments. The economic performance of the former *kolkhozi* and *sovkhozi*, the former collective and large-scale farms, deteriorated significantly. In the crop sector, these farms continue to be the major producers of cereals, sugar beet, and oilseeds, while the production of potatoes, vegetables, and fruits was widely shifted to small-scale agriculture. In the livestock sector, the average large-scale production unit had even greater difficulties, and production in many of the former large-scale units was discontinued. The profitability of both large-scale crop and livestock farms declined, and only 'soft budget constraints' exercised by regional administrations kept many of these farms alive.

Parallel to the decline of the large-scale sector, a *shift in agricultural production* has taken place towards the *private subsidiary plots* (in Russian they are called Lichnie Podsobnie Khozyaistva/LPH), the small-scale production units of the rural populace. By 1998, about half of Russia's agricultural production was produced by this sector. Nevertheless, total agricultural output fell significantly in the 1990s in

1

the Russian Federation. Particularly cereal production plummeted to merely 50% of the production level in 1990. While over-reporting during the socialist era and under-reporting from the beginning of the 1990s on is likely to be one factor which contributed to this sharp decline, the livestock sector in particular suffered from the significant output decline in cereal production. Imports of cereals which comprised the major share in total agricultural imports in the mid–1980s declined, livestock and processed food products were imported in ever increasing quantities. As a result of the significant amounts of processed food commodities that were imported, not only the share of food imports in total imports increased but also the country's agro-food trade balance became increasingly negative. Finally, it is important to note that agriculture used to be one of the sectors in the Former Soviet Union which benefited from the highest preferential treatment compared to all other sectors in the economy. The producer subsidy equivalents (PSEs) as calculated by the OECD for the mid–1980s document that the major share of agricultural producers' revenues were accounted for by state intervention. This situation changed dramatically early on in the 1990s when the increasing government budget deficit no longer allowed subsidization of the large-scale agricultural enterprises at such an unprecedentedly high level and market support diminished due to the abolishment of border protection (see e.g. SEROVA 2000 for more details).

Table 1-1   Russia's economy in the transition period[*]

	1990	**1993**	**1996**	**1998**
GDP (index 1990=100)[1)]	100	74	59	57
Inflation (end year changes in CPI) in %[2)]	n/a	840	22	84
Real unemployment rate, annual average[3)]	n/a	6	9	12
Trade balance in US$, in billions	n/a	16	20	14
Overall government budget deficit in % of GDP[4)]	n/a	10	6	5
Average per capita income in US$[5)]	n/a	4,930	4,190	6,180

Notes: * Most figures are official estimates which do not always take into account informal economic activities.
Sources: 1) Calculated on data base of OECD 1999a. 2) OECD 1999b. 3) GOSKOMSTAT 1998b. 4) Based on WORLD BANK 1999 and 2000. 5) Purchasing Power Estimates of GNP per capita by WORLD BANK 1999 and 2000.

These few stylized facts are indicative of the state of the art of Russia's agricultural sector in the 1990s and highlight the complex interrelationship between sector-specific and economy-wide developments. Generally, the economic performance of the sector changed to some extent since the financial crisis in 1998. The significant real devaluation of the ruble has indeed opened a "window of opportunities" and increased the competitiveness of Russia's domestic food industry and the agricultural sector (SEROVA et al. 1999). This enhanced competitiveness has also resulted in new actors entering agricultural production: large-scale agro-holdings with capital from big investment conglomerates, modern machinery and technology, and western management have contributed at least in some regions to a turn around of agricultural development. However, in most rural areas of Russia market distortions continue to persist, and market efficiency is still low. Against this background, the Russian government is under high pressure to come up with new agricultural policies which will address two major problems: on the one hand transition-specific problems such as the on-going restructuring of Russia's farms, the revitalization of agricultural production and thereby kick-starting the rural economy of the country's vast hinterland. On the other hand, a forward-looking strategy has to be developed to fully integrate Russia's agricultural sector and its food industries in a globalized world economy.

Economic analyses can help to rationalize the choice of policies to support the development of these sectors. While some of the related analytical issues are clearly sector-specific and, therefore, could be addressed in a partial equilibrium framework others are a matter of economy-wide developments. Because of the relative importance of agriculture and the food industry for Russia's economy, its significant inter-linkages with other economic sectors, and its importance for the development of Russia's countryside, the objective of this study is the following: to develop an applied computable general equilibrium model of Russia's economy which pays special attention to agriculture and the food-industries and which reflects some of the most important features of Russia's economy in transition.

General equilibrium aspects of Russia's transition process

From the stylized facts about Russia's first transition decade as they have been outlined above, the following central research questions arise: first, can the links between the economy-wide developments and the development of Russia's agriculture and its food industry in the transition process be analyzed with an applied general equilibrium framework which is both, theoretically and empirically consistent? Second, how can the weak economic performance of Russia's agro-food sector in the 1990s be explained and what are potential ways out of this situation? Third, what are the effects of various foreign trade strategies of the Russian Federation and how will they affect the future development of the country's agro-food sectors? In the following section, we will elaborate the motivation for choosing an applied general equilibrium model further.

The characterization of Russia's transition path in the previous section indicates that this process has been more complex than anyone had expected beforehand. The first period of reforms was preoccupied with liberalization, privatization, restructuring, and macroeconomic stabilization. Institutional reforms were neglected or at least were not at the forefront of the political agenda. There is a widespread consensus that this *lack of efficiency-increasing institutional change* is responsible not only for the slowness of economic recovery in transition economies in general (e.g. JOHNSON 1998; HAGEDORN 1998) but also for the poor state of the economy in Russia in particular (e.g. SHLEIFER and TREISMAN 1998).

Table 1-2　Russia's agro-food sector in the transition period[a]

	1985	1990	1993	1996	1998
Share of agriculture in GDP, in %[1]	9[a]	16	8	9	7
Agriculture's share in investments, in %[2]	15	16	8	3	3
Profitability of Russian farms, in % of all farms[2]	n/a	97	90	21	19[c]
Grain production, in million t[1)2]	99	117	99	69	48
Mineral fertilizer used, kg/ha of sowed land[2]	85	88	46	17	18[c]
Farm land used by type of user, million ha[2]					
by agricultural enterprises	215	210	175	170	166[c]
by private 'family' farms	n/a	0	10	11	12[c]
by household plots and private gardens	4	4	9	10	10[c]
Output by type of user, in % of total output[2]					
by agricultural enterprises	77	74	57	51	50[c]
by private 'family' farms	n/a	n/a	3	2	2[c]
by private subsidiary plots and private gardens	23	26	40	47	48[c]
Share of food in total imports, in %[3)4]	n/a	20	22	25	26
Agro-food trade balance, in $US bn[3)4]	n/a	n/a	–4.3	–8.4	n/a
Average meat consumption, in kg per capita[1) 2]	67	75	59	51	48
Average PSE, in %[1]	81[b]	69	–22	26	19

Notes:　a) Most figures are official estimates which do not always take into account informal economic activities. b) Value for 1980. c) Value for 1986. d) Value for 1997. d) Negative PSE indicate taxation, positive PSE subsidization of producers. However, estimates of this indicator for transition economies have to be interpreted with particular caution. They rely on official data and the official exchange rate, to which they are extremely sensitive. In the case of Russia, on top of macroeconomic and sectoral policies, institutional developments have been decisive for the trend in both measures (see MELYUKHINA, QAIM and WEHRHEIM 1998).

Sources:　1) OECD 1999a and 2000a. 2) GOSKOMSTAT 1998a. 3) GOSKOMSTAT 1998b. 4) GOSKOMSTAT 1999. 5) WORLD BANK 1998.

While *accepting in this study the importance of institutional factors* in explaining the poor economic performance of Russia in the transition period, we address them explicitly in a few sections only. Instead, the focus of this study is the *general equilibrium repercussions of changing economic conditions within the transition period* which affect the development of the agro-food sector. We argue that such general equilibrium effects have a high explanatory power in analyzing the performance of Russia's agro-food sectors in the transition process. Numerous country studies have been presented during the last two decades which used such a general equilibrium approach for analyzing the role of the agro-food sector. The number of such studies for transition economies is still low and until now we have had no such study which provided a consistent theoretical and empirical general equilibrium framework for the post-communist Russian Federation.

Several arguments can be put forward in favor of this approach. First, the transition from plan to market has obviously induced *significant changes in the relative prices* on all levels of the economy. For instance, the immediate effects of the financial crisis which started in mid–1998 resulted in drastic reductions of Russia's agro-food imports, indicating a significant responsiveness of Russian consumers to relative prices changes. In fact, the crisis and the devaluation of the ruble shifted the economy from an unsustainable equilibrium to an equilibrium that will hopefully prove to enhance economic growth. Therefore, the emerging recovery of Russia's agro-food sector in 1999 indicates that *not only a stable macroeconomy* is important but that a second precondition for economic recovery is that *macroeconomic 'balances' are maintained.*

Second, a general equilibrium framework can help to *identify sector-specific policies and economy-wide effects* of general economic and policy changes that induce relative price changes in a specific sector. Indeed, the effects of import substitution policies and other 'indirect' policies effectively discriminating against export-oriented sectors such as agriculture have featured prominently in the debate on Latin America's economic policies in the 1980s and have contributed to improving the economic performance of agriculture in that region significantly. For instance, KRUEGER, SCHIFF, and VALDÉS (1991) and WIEBELT et al. (1992) have highlighted that the level of agricultural sector protection may be significantly influenced by economy-wide distortions such as distorted terms of trade. In this context, the general equilibrium framework should be understood as an important didactic tool to better understand the economy-wide effects of sectoral policies as well as the sectoral implications of macroeconomic shocks. Hence, it would be a mistake to neglect the importance of such macroeconomic developments when analyzing the transition economies.

Third, while recognizing the importance of *institutional change* for the development of Russia's agro-food sector (e.g. WEHRHEIM 1998), the respective effects are taken into account in this general equilibrium framework only indirectly. In a similar way to neoclassical theory, *new institutional economics* employs price theory as an essential part of the analysis (NORTH 1995: 19). In fact, one of the major implications of institutional change not only in transition economies is its *impact on relative prices*. We will argue that institutional change in the transition period has been insufficient to create a fully flexible price mechanism that corresponds with the neo-classical result of fully efficient markets; this will be revealed by modifying the standard general equilibrium model in such a way that it better reflects some of the important characteristics of the Russian economy in the mid–1990s. In fact, one critique of the ongoing transition process in Russia is that, in principle, prices have been freed, but the development of the institutional framework which is needed to grease the price mechanism (in German referred to as *Ordnungspolitik*) is lagging far behind. Many market imperfections remain, and these, in principle, result in a *'sticky' price mechanism* that can be accommodated in the analytical framework of a general equilibrium model. We argue that many *institutional factors affect the economy via the price mechanism*. For instance, poor legislation for farm privatization can reduce the motivation of workers in large-scale farms. This again would contribute to a deterioration in sectoral productivity, which is likely to reduce the relative competitiveness of the sector within the economy and therefore should induce restructuring within the economy. In this study, we do not attribute such changes to specific institutional changes but instead assume that they are exogenously determined.

Fourth, as the Russian economy makes further progress on its journey from plan to market, the Government of Russia urgently demands analytical tools for policy analysis and monitoring (GOVERNMENT OF THE RUSSIA FEDERATION 2000: 6). In this context the *general equilibrium framework* will become an increasingly important *instrument in the toolbox of policy analysts*. However, it should be considered as a tool to facilitate the understanding of the economy-wide effects of general economic change in the Russian economy rather than a tool for providing quick answers to open and complex policy questions.

Due to the complexity of economic developments in the transition period, it is also evident that we will not model each behavioral peculiarity of the whole economy but instead have to focus on selected aspects of economic transition. This particularly refers to the fact that we have to rely on official data and therefore have to restrict the analysis to the *official economy*. In contrast, the rapidly emerging and restructuring *hidden economy* is much more difficult to include in such a model framework and is therefore outside the scope of this study. However, an attempt is made to address the *important role of small-scale agricultural production*, namely household production, which in the end of the 1990s constitutes a sizable part of Russia's *informal economy*.

Focus on the agro-food sector

In general, *special attention is given in this study to Russia's agro-food sector* and its role in the transition period. There are various reasons for selecting this sector:

Theoretically unexpected mal-performance. There is an unexpected mal-performance of this sector which has not been predicted *on theoretical grounds*. KOESTER (1998) pointed out that at the outset of reforms in the early 1990s many professional economists believed that the introduction of market mechanisms in the transition economies would quickly yield growth and improvement in overall welfare. It was expected that the successor states of the Former Soviet Union (FSU) would quickly resume the exploitation of their comparative advantage in agricultural production, which many westerners attributed particularly to Russia. In fact, Russia was the world's largest exporter of cereals before 1914. However, the fact that the country could not keep pace with the expectations is particularly alarming with respect to the agro-food sector. This begs the question: why has the transition from a planned economy to a market economy not yet yielded higher efficiency and output in agricultural production? Indeed, Russia's economy performed generally worse than the transition economies in Central and Eastern European Countries (CEEC) for a long period, despite the fact that initial conditions were relatively similar.

Size of the country. Another good reason for choosing the country as the focus of the study is Russia's size and its unprecedented wealth of natural resources. The performance of the country's agro-food sector is therefore also of interest with respect to world agricultural markets. Additionally, even though the problems vary to some extent between Russia and other countries of the FSU, it nicely exemplifies the problems prevalent in the agro-food sector of the whole region.

Role of agriculture in the economy. The agro-food sector of Russia has been chosen because agriculture continues to play a vital role in the whole economy in spite of the tremendous decline in output experienced in the first transition decade. It continues to be important both from the production and even more so from the consumption side. Furthermore, the agro-food sector has strong backward linkages because of its purchases of intermediates and also has forward linkages by selling its output to other sectors in the economy and to final consumers. The agro-food sector demands labor and capital and contributes to household income. In fact, food expenditure shares have risen to substantial levels during transition, which means that the distributional effects of changes in the agro-food sector are likely to be substantial. Additionally, due to its firm cultural and historic rooting, the sector's development also has widespread social and political ramifications. In fact, many of the economic problems Russia faces today can be exemplified by analyzing the situation of the agro-food economy in Russia.

Policy advice is demanded. Finally, because of the evident problems Russia's agro-food sector is facing, policy advice based on economy-wide modeling is asked for (TACIS-SIAFT project proposal 1999). On the one hand, domestic issues such as the budgetary effects of agricultural policies necessitate quantitative assessments of various policy options. On the other hand, there is an increasing integration of Russia's agro-food sector into global markets, and it is likely to join the World Trade Organization (WTO) within the near future. In this context, the wide variety of different policies and their sector and economy-wide effects ought to be analyzed quantitatively. We will show that a general equilibrium model for the Russian economy can be a powerful tool for addressing such issues.

Objectives of the study

As mentioned above, it is the major objective of this study to develop an applied computable general equilibrium (ACGE or CGE) model for the Russian economy in transition and make its theoretical and empirical components as transparent as possible. Making the 'interiors' of the model public enables a better understanding of the results, but also means it can be used as a benchmark in future efforts to model the Russian economy. The second major objective of the study is to use the model for a quantitative assessment of issues related to the second and third research question posed in the previous section. Specifically, we want to assess the allocative effects of two types of experiments: first, various policy experiments on *specific sectoral developments* related to primary agricultural production and/or the food industries. Second, it will be used to look into the effects of *sector-neutral economic events* such as changes in macroeconomic conditions and their repercussions in the agro-food sector. Research question one will be addressed in the process, by discussing the potential usefulness and the obstacles of general equilibrium analysis of the transition process in Russia.

Notwithstanding the high demand for general equilibrium analysis of Russia's economy in transition, there are *still many obstacles to carry CGE-based analysis in the context of the Russian economy*. First, and most importantly, the deep causes of the above-mentioned institutional impediments to better functioning markets have not yet been studied in sufficient detail to be incorporated more explicitly in a general equilibrium framework. Second, data constraints arise, particularly because of poor representation of the shadow economy in official data sources. Neither of these arguments should prevent the use of a general equilibrium framework but at the same time they indicate the need for great caution when interpreting results of respective analyses.

Outline of the study

In chapter 2, we start out with a discussion of the *transition process* which has driven the economic development of the Russian Federation in the 1990s. This is important, as it has effects for the modeling framework as well as for the kind of policies which will be analyzed later on. First, it is obvious that the transition process has not yet unleashed all market forces to their full flexibility, which means that the model we develop in this study is more rigid than the standard, neoclassical model. It will contain one disequilibrium feature, incorporate inflexible labor and capital markets and be mainly demand-driven, which means we will *classify it as a structural model*. Some of the major theoretical implications of this model specification for various policy simulations and exogenous economic shocks will also be discussed in chapter 2. Second, we will categorize the factors which have the potential to affect the development of the agro-food sector in the transition period and identify those effects which can be addressed with a comparative-static general equilibrium model.

The theoretical structure, the equations of the CGE model, and the respective modifications that turn it into a structural model will be discussed in chapter 3. The general procedure in developing the data base for an ACGE model, as well as the specific problems encountered in creating a consistent empirical version of such a model for Russia's economy in the 1990s, are elaborated on in chapter 4. The steps which were needed to achieve a fairly far-reaching disaggregation of Russia's economy (a total of 20 sectors are distinct from each other) and particularly of the agro-food sector (10 out of the 20 sectors in the economy belong to the agro-food sector) are also discussed in this chapter. Furthermore, in this chapter, we describe in detail how the data base was updated from 1990 to 1994 because it will be an important task to continuously update the data base of such models in the future as better data becomes available. In chapter 5, various policy and economic experiments are presented to discuss major links between Russia's agro-food sectors and the rest of the economy in the transition period. Sensitivity tests will be carried out and the model will be validated by asking if the effects of specific exogenous shocks actually replicate real world developments which were experienced in the transition period. In the concluding section, chapter 6, research and, more cautiously, policy conclusions based on the simulations will be drawn.

Chapter 2

Empirical and Theoretical Considerations on Russia's Transition Process and Related Implications for the General Equilibrium Analysis

The objective of the following chapter is twofold: first, we conceptualize the transition process and discuss its impact on the Russian economy in general and on the agricultural sector in particular. A cross-country comparison will indicate where Russia stands in the group of transition economies. This comparison will indicate that the development of the economy as a whole and the general investment climate in a specific transition economy have been quite influential for the development for agriculture. We proceed by discussing how the transition period has evolved specifically in Russia and to what extent it has been conducive to creating functioning markets. The second major objective of this chapter is to identify the implications of the country-specific transition experience of Russia for the proposed means of analysis, i.e. a general equilibrium framework. We will argue that a purely neoclassical model will not be appropriate under the given conditions of the Russian economy and therefore discuss the features of a modified general equilibrium model and outline the economic causality within such an alternative model specification.

Conceptualizing the transition process

The term *transition* refers to the *set of reforms and the timing and sequencing of reforms* with which a country implements the *system switch from plan to market*. This move from a centrally planned economy to a market economy based on decentral market coordination followed rather different patterns in the respective countries. To understand today's diversity in the economic systems and performance of the transition economies, it is helpful to conceptualize this process (Figure 2-1).

On the eve of transition, the *initial conditions in the transition economies* were different and from that time on the transition path chosen became increasingly country-specific. Three major dimensions of the transition process explain these ever increasing differences: first, due to different political forces at work, as well as cultural characteristics and social conditions, the *initiated reform process* soon became country-specific; second, *the mechanisms* that were set in motion by these reforms are characteristic for each country; third, *the institutions* which create the framework for a market economy and which have emerged from these overlapping processes are very different. The transition process, which was by no means completed at the turn of the millennium, thus resulted in rather unique country-specific institutions which are decisive for the extent to which the current conditions reflect decentral market coordination. We will elaborate this scheme in more detail in the following.

On the eve of transition, the common and economically decisive characteristic of the respective countries was that they were centrally planned economies, though to differing degrees and with very different endowment of production resources. Hence, it is important to acknowledge that the *initial conditions* on the eve of transition have turned out to be crucial for the economic results of this process. More specifically, MACOURS and SWINNEN (2000) showed that the initial conditions in transition economies have had significant effects on the economic development of agriculture in the course of transition.

It is obvious that the *endowment of the respective country with natural as well as with human resources* is essential as they define the production possibility frontier of the individual country. Human capital is important in terms of technical and intellectual knowledge, as is social capital in terms of prevalent norms, traditions, and attitudes of the populace. It is therefore also evident that 'non-economic' factors such as the length of the centrally-planed era, which has had profound effects on the human capital factor, are of pivotal importance.

Second, the country-specific approach to economic reform mattered. The reform patterns were different with respect to the *type of reforms*. In the CEECs, macroeconomic reforms and market-oriented institutional reforms were implemented in parallel. In Russia, macroeconomic reforms were implemented rather instantaneously, rather like a *big bang*, and the development of urgently needed market institutions lagged far behind. Hence, it is evident that the *timing and sequencing of reforms* varied significantly between transition economies. However, it is important to note that the initial conditions also influenced the choice of reforms to some extent. The geographical differences between the CEECs and the Russian Federation, for instance, explain why Russia was inclined to invest a great deal of time in the decentralization of policy-making, which had significant effects on the mechanisms set in motion.

In fact, the choice of different macroeconomic reform policies further illustrates the diversity of reform approaches. The *four large areas of macroeconomic reforms*, namely liberalization, privatization, restructuring, and stabilization, were initiated in most countries, though the timing and sequencing has again been unique to each country. While these reform packages were designed to be *sector-neutral* at the beginning of the reforms, the demands of political interest groups and the economic characteristics of individual sectors resulted in *sector-specific* reforms. The best case in point is agriculture, for which in many transition economies (with notable exceptions such as Estonia or Kyrgyztan) the initial liberalization was reversed to once again grant domestic producers increased preferential treatment in the course of transition.

In very few countries, the fundamental reforms have not been implemented at all, a case in point being Belarus. In other countries, the implementation of such macroeconomic reforms has been more gradual, for example in China. While the long-term growth perspectives are rather blunt for Belarus because these reforms

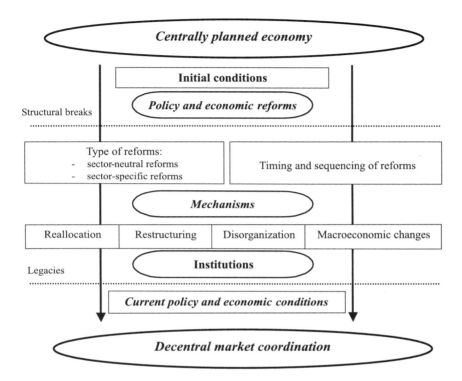

Figure 2-1   The (country-specific) process of transition

have not been implemented at all, they are rather optimistic for China. Hence, it is evident that both *the mix and the timing and sequencing of reforms has had a significant impact on the economic growth path of the transition countries.* Generally the 'quality' in the timing and sequencing of reforms should materialize in better economic performance. The more efficient the mix and the timing and sequencing of economic reforms, the more quickly the risk of investing in the particular country should diminish; the more successful the implementation of reforms, the more quickly economic growth would be expected. Generally, it seems as if the longer the transition period lasts, the more country-specific the reform path becomes.

Notwithstanding, the combination of sector-neutral and sector-specific reforms sets *similar economic mechanisms in motion,* even though at very different rates. BLANCHARD (1998) distinguished *four such basic mechanisms* that were said to drive the economic transition process, at least in the initial transition period: reallocation of resources, disorganization of firms, restructuring of the economy, and macroeconomic policies. The very purpose of these mechanisms is to dismantle the *legacies* of the socialist era that have been crucial elements of the central market coordination. The two related mechanisms of reallocation (of resources) and disorganization (of the formerly state-controlled large-scale enterprises) were initially associated with higher unemployment. The restructuring of the transition economies can explain why at least the transition process in the CEECs was associated with increases in productivity and only limited gains in employment. With respect to macroeconomic policies, Blanchard gives equal weight to two policy objectives: on the one hand, shifts in aggregate demand should not be prevented by excessively tight monetary policies; on the other, macroeconomic stabilization plays a key role in output recovery. Blanchard argues that these four *basic mechanisms* were initiated in the early years of transition in almost all transition economies and have yielded in the respective countries a rather characteristic, common result: a U-shaped response of GDP was visible in those transition economies which implemented the relevant reforms.

In spite of this common economic response, *the sectoral effects of these mechanisms were rather different.* A good case in point is again *agriculture.* While the mandatory reforms to start, for instance, farm restructuring have been implemented in most CEECs and most economies of the FSU, the effects were rather different. The extent to which the restructuring in the farm sector has actually taken place seems contingent not only on the country-specific reforms but also on the initial attitudes and perceptions of the rural populace with respect to private ownership and entrepreneurship. While on the eve of transition a collective memory of "What markets are like" still existed in the CEECs, it had widely faded away in the FSU. Therefore the mechanisms which were set in motion by the initial reforms yielded rather different results, which also affected the shaping of new institutions. Due to the differences in the reform paths of individual transition countries, a set of country-specific institutions has evolved. As a result, the extent to which decentral market coordination takes place today is also very different.

The variance in these factors is likely to explain many of the differences in economic outcomes of the first transition decade. It also seems likely that they are essential for the performance and economic development of agriculture. In the following section, we will use a cross-country comparison of transition economies to support this argument with some empirical observations. We will look at the differences in general economic performance and the agricultural performance achieved in some selected transition economies in the initial transition decade.

Cross-country comparison of economic performance in transition economies at the end of the first transition decade

Generally the CEECs and the countries of the FSU have chosen different reform paths and also started from very different initial conditions. In an attempt to evaluate the success of reforms, it might be tempting to look first at output measures such as GDP and Gross Agricultural Output (GAO). However, as discussed in the previous section, initial conditions may have had significant effects on the output performance.

While there is no other readily available measure regarding the 'quality' of reforms, an implicit indicator for the progress of reforms is *the risk of investing in a particular transition economy.* The country-specific 'risk', particularly of emerging market economies, has been judged by many international rating agencies. One risk indicator that is available for most transition economies is the Euro-money Country Creditworthiness Index (ECCI). In Figure 2-2, the country-specific value of this index is shown for 1997, the first year in the transition period for which it was available for most countries in the region. The ECCI ranges here from 0 (lowest risk) to –20 (highest risk). The benchmark which defines the zero risk level is the United States of America.[1] The ECCI (left bar in Figure 2-2) reveals significant variation for the countries in this sample, indicating that the reforms of the initial transition period have yielded very different results.

In order to get a *more broadly based risk indicator*, the data for the ECCI has been combined with other risk indices (right bar in Figure 2-2). Based on three risk indices that have been compiled by different rating agencies, the transition economies of the FSU and the CEEC were ranked according to their respective country risk.

Both risk indicators revealed in Figure 2-2 imply that there is a *geographical divide in the level of risk moving in an eastward direction* in the region. The Czech Republic, Poland, Hungary, and Slovenia, the most western countries in this group, reach the highest rank, indicating that by 1997 it was least risky, at least within this group, to invest in these economies. At the same time it implies that these countries have made the most significant progress with respect to implementing market-oriented reforms since the beginning of the transition process. In contrast, the countries of the FSU are to be found on the right side of the diagram, indicating lower ranks associated with higher risk levels. The better risk performance of the CEECs implies that the reform path chosen by these countries has been superior to that chosen by the successor states of the FSU. This fact in itself is not yet very informative. The question that is more interesting in the context of this study is whether the differences in the country-risk level are associated with different economic results. Or, putting it differently, we ask if the degree to which economic reforms in the transition economies have yielded lower risk levels goes hand in hand with higher output growth in the whole economy as well as in agriculture.

Given the differences in reform attitudes and the differences in initial conditions, it is not surprising that GDP and GAO have, in fact, developed very differently in the countries of this region in the first decade of the transition period (Figure 2-3).[2] Note that the countries are in the same order as they were ranked in Figure 2-2 according to their risk-level. The *annual growth rate of GDP for the first transition decade* indicates that there is some kind of *a west-east divide*. While average annual GDP has even risen over the observed period or declined only slightly in the CEEC countries, it decreased substantially in most countries of the FSU.[3]

The second country bar in Figure 2-3 shows the average annual growth rate *of gross agricultural output* between 1990 and 1998. This indicator reveals that, after a

---

[1] See also footnote to Figure 2-2 for further explanations on this index.

[2] For some countries out of both regions, no reliable data was available due to civil war or other disruptions. Hence, they were omitted.

[3] In another article WEHRHEIM (2000b) showed that this west-east pattern can also be observed with respect to poverty: the risk of being poor increases the further east one lives in the region.

decade of transition, the negative effects on agriculture have been reversed in some CEECs, or at least were more moderate compared to the countries of the FSU. In some countries of the FSU (Armenia, Azerbaijan, and the Kyrgyz Republic), the output decline in agriculture was far less pronounced – or even positive – compared to overall GDP, and might even have been a buffer against more significant drops in overall national output in this period.

Therefore, the *geographical pattern of agricultural development is less distinct* if compared to the general economic development of transition countries. Annual average growth rates of GDP and GAO between 1990–98 were positive only in three countries, with Slovenia being the only country that achieved growth of both GDP

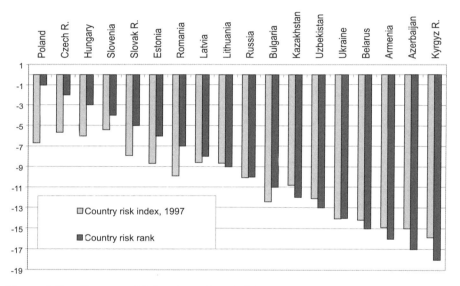

Figure 2-2   Country risk levels and rank[a] for selected transition economies for 1997

Notes: a) The Euromoney Country Creditworthiness Index indicates the 'risk level of investing in an economy'. This index is reported for all countries, but no information is available on the number of risk components on which this index is based. For illustrative reasons, it has been converted and ranges here from 0 (lowest risk) to –20 (highest risk). However, other risk indices are available and yield different evaluations. Therefore, a country risk rank has been compiled by calculating the average risk rank for each country based on the ECCI and two other indices: 1) The Composite International Country Risk Guide is based on 22 risk components; it is reported, however, only for a few countries; 2) Institutional Investor Credit Ranking indicates the risk of a country's default. A high rank indicates a low level of country risk. The negative ranks are shown for illustrative reasons. The country risk rank is based on information for 1997, as no risk indicators were available for previous years.

Source: Data from WORLD BANK 1998.

and GAO. Furthermore, it is interesting to note that agricultural sector development has been decoupled from general economic development in only a few countries. This stylized observation suggests that *sector-neutral factors have been rather important for the development of agriculture* in most transition economies. This is so in spite of the fact that endowment with natural resources is normally expected to be the most decisive factor for shaping the comparative advantage of agriculture.

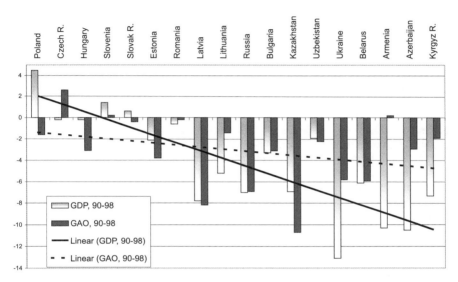

Figure 2-3   Annual average growth rate of Gross Agricultural Output/GAO and of Gross Domestic Product/GDP for 1990–98 for transition economies

Source: Data from WORLD BANK 1998.

Finally, this cross-country comparison indicated that *Russia was not the most successful country in pursuing the shift from plan to market*: the development of GDP and GAO within the 1990s has been equally disappointing. Does this imply that the same factors that drove general economic development also affected agricultural sector performance in Russia? In the following, we will therefore look at the country-specific economic situation and discuss some of the economy-wide trends which are likely to have affected agriculture and which, at the same time, have implications for the development of a general equilibrium model.

Selected economy-wide trends in Russia's transition process

For almost a whole decade, a coherent concept for timing and sequencing of reforms was lacking in the Russian Federation. Political struggles between federal and regional forces, between the executive and the legislative bodies, and between political interests groups and economic oligarchs prevented the development of a stronger consensus on the path of transition. This indeed seems to be part of the explanation of why the Russian transition process has yielded limited positive effects both in the general economic situation and in agriculture. In addition, the OECD (1997) argued that the bleak record of the early stage of the Russian transition reflected the difficulty of overcoming the dismal legacy of seven decades of Soviet power. Two dimensions of the reform process that were initiated early on and more or less at the same time in Russia's transition process are good examples for policy developments which have not always been harmonized but, nevertheless, had economy-wide effects: liberalization and decentralization.

Liberalization. In the first few years of the transition process, liberalization in Russia took place in its *foreign and domestic markets.* The foreign trade regime of the Former Soviet Union, in which the state was the only actor, was abolished; by 1992, Russia had a fairly liberal trade regime (OECD 1999a). Not only import tariffs and export constraints were widely dismantled but also the exchange rate and the foreign exchange systems were liberalized. In fact, black markets for currency exchange had widely disappeared in 1993. Even though the inflow of imports was substantial because of even larger raw material exports, Russia ran a trade surplus. By the mid–1990s, Russia's economy was in fact a very open economy, which had rather negative implications when the world financial crisis which moved around the world started to affect the positive economic development in Russia at the end of 1997 and culminated in the financial crash of the ruble and the financial sector after mid–1998. The lack of an adequate institutional framework, which in western economies helps to reduce market failures, is an additional factor determining why the world financial crisis hit Russia much harder than, for instance, the CEECs.

The second major platform on which liberalization became effective early on in Russia's transition process was domestic markets. Domestic prices were liberalized in 1991 and 1992, and at the end of 1993 consumer and producer prices for most commodities were set free and formed in response to market forces. Enormous numbers of traders emerged in Russia early on in the 1990s, and by the mid–1990s a competitive market environment had developed, particularly in the market for consumer goods. This also helped to reduce inflation, which in 1994 was much lower than in the years before.

Decentralization. Ever since *Russia became a Federation with 89 subnational subjects* (i.e. oblasts, krais, autonomous republics, and the two metropolitan areas, Moscow and St. Petersburg), *decentralization* introduced various additional layers of policy-making. This impinged on the transition process because of diverging regional interests, which again had strong impacts on the objectives and the choice of reforms implemented in the transition process at various levels of policy-making in Russia. *Decentralization of agricultural policies* is a good case in point, as many irrational and uncoordinated policies were implemented on the regional level, resulting in a regionalization of agricultural policy-making in the first half of the 1990s (MELYUKHINA and WEHRHEIM 1996). This trend persisted in the second half of the 1990s and at the end of the decade around 75% of total budgetary expenditures for agricultural support were made by regional governments (SEROVA and KHRAMOVA 2000b).

The *economic consequences of decentralization* for agricultural markets were mostly negative: it contributed, for instance, to strategies of regional self-sufficiency in food production with its *detrimental effects on spatial food market integration.* Particularly with respect to retail markets for food commodities, various authors have shown that the degree of market integration was rather low, at least as compared to western economies (e.g. GARDNER and BROOKS 1994; LOY and WEHRHEIM 1999; GOODWIN, GRENNES, and McCURDY 1999).

KUHN (2000) argued that not only regional trade barriers but also *high risk and generally high transaction costs associated with long-distance trading* in Russia are among the detrimental effects of decentralization that prevent better functioning markets. Furthermore, *regional interests have prevented more significant restructuring* of local and regional economies. In order to keep regional employment records at acceptable levels, and not to reduce the output potential in the primary and secondary sector, many regional governors have prevented that large-scale industrial plants as well as large-scale agricultural enterprises were shut down. Such *soft budget constraints* implemented by regional governments have helped to keep many large-scale agricultural enterprises alive. This contributed to the observation that in the late 1990s between 80% and 90% of Russia's former collective farms were operating without (officially) making profits (e.g. IANBYKH 2000; ZEDDIES 2000).

One other frequently quoted negative economic effect of decentralization is the *formation of regional monopolies.* In fact, there have been indications that regional monopolies have formed, particularly in selected and strategically important segments of the food industries. At the same time, the competition legislation in Russia primarily assumed the form of anti-monopoly policy and to a much lesser extent a pro-competition policy. However, the emergence of small-scale actors in many food industries, such as the meat industry, has by and large prevented the emergence of persistent monopolies in the 1990s (NUPPENAU and WEHRHEIM 1999). WANDEL (2000) argues that imperfect competition in Russia's agro-food sector was still rare at the end of the 1990s but care should be taken that some forms of vertical cooperation and integration do not result in a restriction of competition. Particularly on a fairly high level of aggregation, in terms of industries and spatially, there is no evidence that Russia's markets did not behave competitively in the mid–1990s.

One of the first policy measures initiated by President Putin in the summer of 2000 was to *reverse the trend towards greater decentralization* of decision-making. The motivation to reduce the devolution of power might have been dominated by arguments of political power. At the same time, it might prove to be an urgent reform to further reduce some of the most effective market failures and structural rigidities within the Russian economy.

These coinciding trends of decentralization and liberalization have various implications for our analysis. On the one hand, the various components of liberalization have helped that the 'invisible hand' started to rule in most markets in Russia early on in the transition process. Competitive price formation is the rule and not the exception. However, against the background of Russia's vast territory, the process of decentralization has been one major obstacle to creating smoothly functioning markets which would be free of market distortions. Low food market integration and stickiness of capital investments are examples of the kind of rigidities which the Russian economy is faced with today. While *we will not look into the specific economic effects of decentralization within this study*, we will discuss in the next section how some of the prevalent rigidities in Russia's economy can be accommodated for within a general equilibrium framework.

Implications of Russia's transition path for a general equilibrium framework: a structural instead of a purely neoclassical model

A general equilibrium analysis of the economy as a whole rests on the basic assumption that the prices of all goods and services are determined simultaneously by assuring that *all* markets in the economy are in equilibrium. At these equilibrium prices, the supply of each commodity should be equal to demand. In equilibrium, there would be no incentive to change behavior for any economic agent in the system. If the system is exposed to an exogenous shock, relative prices have to adapt until a new equilibrium is found. This also implies that the causality within such a model always starts with price changes which induce quantity changes and not vice versa.

While this condition must hold for most types of general equilibrium models, *the adaptation path to a new equilibrium can alternate*. Depending on the adaptation path, various types of equilibrium models could be identified. In the following, we will make a distinction between two rather different types of models and, for the sake of simplicity, call them *neoclassical* and a *structural general equilibrium model*.

Neoclassical equilibrium model. For many decades, the major paradigm in economics was the neoclassical one, which strongly rests on the assumption that *relative prices* are the central variable in explaining the behavior of markets.[4] The adaptation path in such models rests on the concept of *perfect markets*. Prices can adjust smoothly to a new equilibrium because of various conditions assuring that markets are 'perfect' (e.g. complete information of all economic agents about all relevant variables, perfect competition, homogenous goods etc.). If price flexibility were perfect, any change in relative prices would transmit into a response in quantities of equal dimension. Another assumption of the neoclassical model would be that investments are determined by savings. The national interest rate would be an important variable by assuring that aggregate demand always equals national income. Therefore, the neoclassical model would be built on the assumption that all markets, for products as well as for production factors, are fully cleared. Generally, the neoclassical model economy rests on the assumption that perfect competition

---

4   While this paradigm is still central to economic thought, it has been extended by the assumption of imperfect markets. In many economic fields, institutional economics and particularly the transaction costs approach pioneered by Coase, Williamson, and North have complemented the neoclassical paradigm and today are integral parts of economic theory. This is true, for instance, for the economic sub-discipline of industrial organization (WEISS 2000: 406) but also for transition economics (WEHRHEIM and VON BRAUN 2000: 528). Because the 'traditional neoclassical' concept has been extended instead of being replaced, a clear-cut distinction between various schools of economic thought is somewhat difficult.

and a high degree of flexibility on the production side mean that resources always migrate towards those sectors which promise the highest profit rates. Therefore, the neoclassical paradigm assumes that supply-side factors determine national production, output and aggregate demand, and not vice versa.

Structural equilibrium model. The structural general equilibrium model we will describe rests on the same pillars as the neoclassical model in that relative prices must force the economic system to an equilibrium. However, it also contains some structural rigidities with which the responsiveness of the model economy is reduced. For instance, while supply-side factors such as technology effects are the driving forces in the neoclassical model, the structural model pays more attention to demand-side factors. Basically, the structural model starts the analysis by looking into the factors which determine aggregate demand. Such an analysis is only useful if the reliability of the neoclassical model assumptions are in question (HENRICHSMEYER, GANS and EVERS 1991: 379).

More specifically, the structural equilibrium model we will use for the analysis of the Russian economy will rest on the following characteristics: first, restrictions on the supply side would imply that the adaptation of production to a change in relative prices is limited. One possibility to include this feature in the model would be to reduce the intersectoral mobility of production factors. The underlying assumption would be that, for instance, investments made in a sugar plant in Rostov can not easily be withdrawn from the unprofitable firm and re-invested in a more profitable sector, e.g. the construction sector in Moscow. As a result of this structural rigidity on the production side, aggregate demand plays a more decisive role in the determination of the equilibrium. Second, the economy can reach a new equilibrium even if labor markets are not cleared and unemployment occurs. This could be incorporated in the model by fixing nominal wage rates, which would mean the level of employment is determined by labor demand. In fact, this feature departs somewhat from the equilibrium assumption in that excess supply of labor is assumed and, in doing so, the notion of a disequilibrium is introduced into the model. Third, the government can influence aggregate demand by means of fiscal policy. Fourth, the link between investments and savings in the structural model economy would rest on the assumption that investments are driven by aggregate demand. Unless savings rates alternate, investments are basically fixed. Fifth, the time horizon for approaching the new equilibrium is more short-run than long-run, as long-run economic developments are contingent on so many factors that predictions become increasingly crucial. This feature of the model would be reflected by choosing such low values of elasticities as to keep substitution between domestic and foreign products of the same product group low. In the following section, we will indicate some of the implications for the economic causality within the two types of models.

Hence, a structural CGE model rests on the same pillars as a neoclassical one but adds specific features of a given economy. These structural peculiarities reflect the extent to which such an economy deviates from the assumption of the perfect market. Because of this combination of elements from different economic fields CHENERY and ROBINSON (1985) characterized such general equilibrium models as being based on *neoclassical structuralism*.

Graphical presentation of differing economic effects of economic policies in the neoclassical and the structural equilibrium model

A graphical representation can illustrate some of the analytical differences of economic events or policies using a structural equilibrium framework compared to a neoclassical model version. One simple and stylized presentation of the latter is based on a programming version of the so-called *1-2-3 general equilibrium model* (cf. DEVARAJAN, LEWIS, and ROBINSON 1994). The model refers to *one country*, the home country, with *two producing sectors*, and *three goods (see Table A2-1 for the equations of the 1-2-3 model)*. The two commodities produced by the economy of the home country are an *export good* (E) and a *good for the home market* (D^s). The third good in the economy is an *imported good* (M). We assume a single consumer who receives all income and who maximizes his or her utility over the domestic and the imported commodity subject to his or her budget constraint. Consumer utility is thus equal to social welfare. We assume that the home country is small and, hence, faces fixed world market prices for traded goods. Therefore, the model uses the concept of *product differentiation* and *imperfect substitutability* both on the demand as well as on the production side. It is also assumed that the country we are looking at has *an open and not a closed economy*. While in a closed economy supply and demand balances would force the commodity markets to an equilibrium, in an open economy supply and demand imbalances would be equated by exports or imports, which would lead the home country to take world market prices as given as long as trade policies are absent. The 1-2-3 model has no money and therefore relative prices determine the product composition. By viewing the 1-2-3 model as a simple programming model it can also be presented graphically. It then can be exposed to exogenous alterations to trace some of the major general equilibrium results (cf. DE MELO and ROBINSON 1989; DEVARAJAN et al. 1994; WIEBELT 1996: 43). Furthermore, we will then change some of the underlying assumptions of the neoclassical 1-2-3 model in such a way that it reveals a structural model world, and we will use that presentation to investigate the effects of an adverse terms-of-trade shock.

In Figure 2-4, we start with the neoclassical version of the 1-2-3 model. An arbitrary *transformation curve* is depicted *in quadrant IV* (south east), reflecting a certain production technology and a given stock of production factors. In equilibrium, the economy produces where the production possibility frontier is tangent to the price ratio (p^d/p^e) and hence determines also the amounts of the good that is produced for the domestic and for the foreign market. All domestically produced goods that are not exported are effectively treated as non-tradables.

In *quadrant I* (north east), the *balance of trade* is shown under the simplifying assumptions that the prices for the export and the import good are both equal to one because of which the slope of the balance-of-trade constraint is a straight line through the origin. Then, for a given level of the exportable produced by the domestic industry the balance-of-trade line determines how much imports can be bought from world markets. Furthermore, it is assumed that the real exchange rate is identical to one divided by the price for the domestic good (1/p^d), and that there is no foreign capital inflow. Based on these assumptions, the balance of trade constraint is a straight line through the origin with the slope of one. For a given level of exports, this line also determines the amount of imports that can be bought from the rest of the world and consumed by domestic consumers. Revenues from exports are the only possibility to finance imports.

Quadrant II (north west) shows the *consumption possibility frontier*, indicating that "the consumer" has the choice between the domestic good (D^d) and the imported good (M). As a result of the balanced trade account and equal world

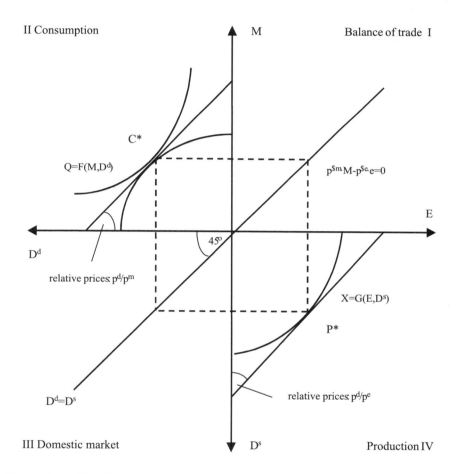

Figure 2-4   Neoclassical version of the general equilibrium model

Notes: The following abbreviations are used: D^d (D^s) = domestic demand (supply); X= total production; Q = total consumption; P* = production of E (the export good) and the D the domestic good in the initial equilibrium; C* = consumption of D (the domestic good) and M (imports) in the initial equilibrium. Pd = domestic price, $p^{\$e}$ and $p^{\$m}$ = prices for exports and imports in foreign currency.

Sources: Adapted from DEVARAJAN, LEWIS and ROBINSON (1994: 2–10) and WIEBELT 1996: 42.

market prices, the consumption possibility frontier is the mirror image of the production possibility frontier. The tangency between the consumption possibility frontier and the indifference curve will determine the split between the domestic and the imported good that is optimum for the consumer.

The former is supplied from the *quadrant III* (south west), which depicts the *domestic market*, in that equal quantities of the domestic good are supplied (D^s) and demanded (D^d) and, hence, are only balanced because of relative prices being equal to one and because no savings are made. What would happen if this neoclassical model-economy were exposed to an *exogenous shock*? Assume that this shock is an *adverse terms-of-trade* shock which could be caused by an increase of import prices.

In this case, the graphical presentation depicted in Figure 2-5 would be relevant: the balance of trade line would rotate outwards and reflect the new price ratio. On the consumer side, the budget constraint adjusts to the rise in import prices such that the consumer has to reduce consumption of the imported commodity. At the new consumption point, the consumer gets less of both, imports and the domestic product. On the contrary, on the production side more exports will be produced in order to be able to pay for more expensive imports. The rise of exportables is due to the fact that the price of the exportables in relation to the domestic good increased, which is identical to a real devaluation.

The change in terms of trade would make exports relatively more advantageous, and the domestic producer would react to this change in relative prices by shifting production resources into the sector producing exports. The exact location of the new equilibrium on both the production and on the demand side depends crucially on the substitution possibilities between the imported and the domestic product.

Generally, a real devaluation (appreciation) would occur if the elasticity of substitution is smaller (larger) than one (WIEBELT 1996: 50). In most empirical studies, elasticities of substitution below one are reported. Furthermore, it is notable that the producer would move along his production possibility frontier to the new equilibrium P1. This is due to the fact that the available production factors in the economy as a whole are assumed to be fixed and respond inelastically to price changes.

At the same time, this implies that all production factors are fully employed and that total output of the economy would stay the same. Hence, this neoclassical model suggests that in the case of any "market disorder" and if the economy is flexible enough to do so, relative prices will adjust in such a way that they will bring the economy into a new equilibrium.

In a *structural model world*, the adaptation to an exogenous shock would initially follow a similar logic: relative price changes cause a reallocation of production and restructuring in demand. However, two major features differ and affect the economic results of the same policy change. First, the reallocation of resources might be constrained, which might imply that the *gravity law does not hold*. This law states that if resources are free to move, they will, c.p. be employed in the sector that yields the highest return. Due to the constraint in resource mobility, aggregate demand will be the driving force. Second, the model might be forced to a new equilibrium, which could also entail some disequilibrium if factor supply becomes perfectly elastic.

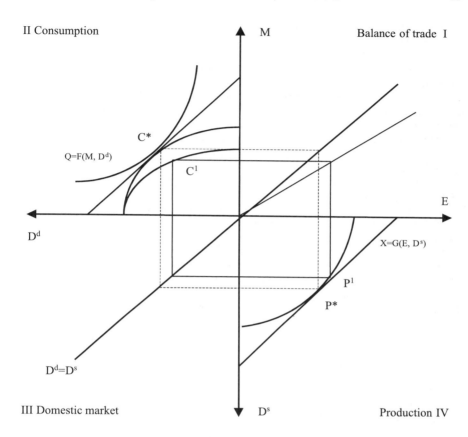

Figure 2-5   An adverse terms-of-trade shock in the neoclassical version of the general equilibrium model

Notes: The following abbreviations are used: Dd (Ds) = domestic demand (supply); X= total production; Q = total consumption; P* = production of E (the export good) and the D the domestic good in the initial equilibrium; C* = consumption of D (the domestic good) and M (imports) in the initial equilibrium. Pd = domestic price, p$e and p$m = prices for exports and imports in foreign currency.

Sources: Adapted from DEVARAJAN, LEWIS, and ROBINSON (1994: 2–10) and WIEBELT 1996: 42.

Perfectly elastic supply of capital is often used in the discussion of the real exchange rate for a small economy (OBSTFELD and ROGOFF 1996). This assumption would not be realistic for Russia, for which capital was constrained throughout the first transition decade. In contrast, the *assumption of perfectly elastic supply of labor* reflects an important feature of Russia's economy in transition, during which official unemployment has been rising and the extent of hidden unemployment has been dismantled by the financial crisis (CHASHNOV 1999). However, the condition of perfectly elastic supply of labor would be different from the classical model that influenced development economics throughout the 1950s and 1960s: the article by LEWIS (1954) on "Economic development with unlimited supplies of labor". In the general equilibrium model proposed by Lewis, it was agriculture that was responsible for the supply curve of labor being infinitely elastic because of a significant wage differential between agriculture and non-agriculture, the capitalist sector. In our analysis, we will assume that the nominal wage rates are equal across sectors. What are the implications of such perfectly elastic labor supply while the level of capital is fixed at its predetermined level and inter-sectoral mobility of capital is impossible? MUNDLAK (2000: 110 ff.) shows this in a fully competitive general equilibrium case in which "... one of the [two] production factors is perfectly elastic, the price is constant, and the product supplies are also perfectly elastic". If the economy were closed, output would be exclusively determined by demand. As we will look at an open economy, trade can impact the product composition to some extent, depending on the elasticities of transformation, but domestic aggregate demand will still remain the decisive force driving the composition of output.

In effect, while the reaction to an exogenous shock on the consumption side might be much the same as in the neoclassical model, the adaptation on the production side might be different in a structural model: *instead of a shift along the transformation curve, an outward or inward shift of the efficiency frontier of the economy might occur* (Figure 2-6).

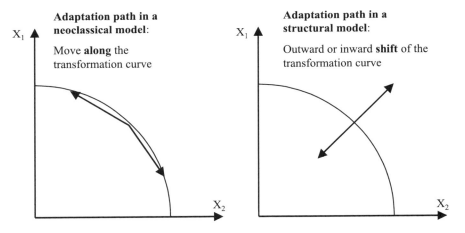

Figure 2-6   Adaptation of production to an exogenous shock in a stylized neoclassical and in a structural model economy

This would also have implications for explaining the *response to an exogenous shock within such a structural general equilibrium framework*, a graphical representation of which is shown below (Figure 2-7). One of the central assumptions of the structural model economy which we will use in the applied version of the model is that of fixed wage rates. This corresponds to the observation that in Russia the wage rate for official employment in the mid 1990s was still determined between the trade unions and major employers. It is only in the informal sector, which we disregard here, that wages are more flexible. By keeping the price for the only production factor – labor – constant, the price of the home-good also remains unchanged.

Now consider an adverse terms-of-trade shock of the same kind as it was analyzed in Figure 2-5, and which again could be due to an increase in the world

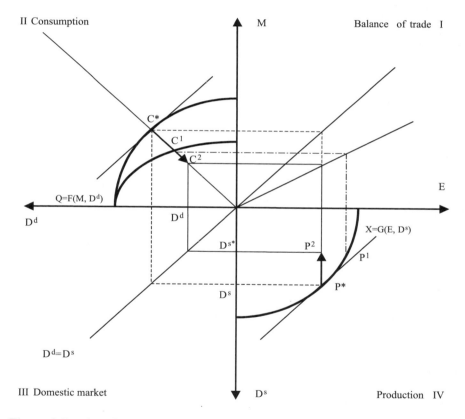

Figure 2-7   An adverse terms-of-trade shock in a structural version of the general equilibrium model

Notes: The following abbreviations are used: Dd (Ds) = domestic demand (supply); X= total production; Q = total consumption; P* = production of E (the export good) and the D the domestic good in the initial equilibrium; C* = consumption of D (the domestic good) and M (imports) in the initial equilibrium. Pd = domestic price, p$e and p$m = prices for exports and imports in foreign currency.

price of the imported good. This would again result in a downward shift of the balance-of-trade constraint in Quadrant I. As the price for exports remains constant, this shift affects the relative prices between the imported and the domestic commodity (P^d/P^m). Therefore, total welfare as measured by consumer utility in the economy will decline because the country can now afford fewer imports. This is depicted in Quadrant II, where the consumption possibility frontier rotates inwards. In the neo-classical model and under a flexible price system, consumption would contract, but the restructuring of demand would result in the new consumption location C^1. The line that links the C^1 with the balance-of-trade-constraint and the production possibility frontier indicates that in the neoclassical model economy the new production location (P^1) would be on the production possibility frontier (compare with Figure 2-5).

In the structural model economy, however, the price of the non-tradable P^d does not change because wages are fixed. As long as the price for exports P^e does not change, there are no incentives to expand the production of exportables, because the real wage rate (w/P^e) remains constant. If exports remain at the previous levels, the available income consumers can spend is only C^2. Under a homothetic utility function, consumers would also reduce the demand for the home-good. As domestic demand for the home good D is smaller than the amount the economy could supply, unemployment occurs: the new production location P^2 is below the production possibility frontier. At this production location, the produced quantity of exports is the same as in the base period, but the quantity of the home-good has declined.

The conclusion from this discussion is that a structural general equilibrium model will take some of the characteristic features of the Russian economy better into account compared with a purely neoclassical equilibrium model. In particular, it incorporates inflexible factor markets and thereby does not rest on the neoclassical assumption of full employment. However, it should be understood that the model we will develop in the next chapter is not built on a clear dichotomy between a neoclasscial and structural models. *Instead, the model will reflect a continuum*: it will rest on some neoclassical assumptions about the behavior of economic agents such as competitive markets and supply and demand balances in commodity markets, but will be 'closed' is such a way that it reveals various features which turn it into a structural model instead. These features will represent various market imperfections which have been mentioned in this section. The technical aspects of this modeling attempt will be discussed in the next chapter (see chapter 3).

The next question which has to be addressed here is which general equilibrium effects we would expect from any exogenous shocks on agriculture. However, we will first look into one other question: To what extent are the general factors driving agricultural development relevant for agriculture in transition, which additional forces might come into play, and to what extent can these factors be analyzed in a modified general equilibrium framework as outlined above?

Conceptualizing the factors affecting the economic development of agriculture in the transition period

While some of the factors affecting agriculture are characteristic for the transition process, it is questionable to what extent these factors were different from the process of agricultural development in the context of economic growth in general. MUNDLAK (2000) empirically identifies the supply-side factors that have affected agricultural development in the past by means of a cross-country comparison. He discusses the theoretical effects of these factors first in a partial and then in a neoclassical general equilibrium framework. Mundlak attributes the greatest importance to three sets of factors affecting agricultural development: production factors (labor and capital), technology, and policy. With respect to agricultural policy analysis of transition economies, two general observations given by MUNDLAK (2000: 57) are particularly relevant: first, "Whatever the motives behind them are, the [agricultural] policies cause distortions and have a welfare cost"; and second, "However, agriculture is not only affected by policies directed toward the sector, or *sector-specific* policies, but also by *sector-neutral* policies, such as macro policies. In fact, often the effect of such sector-neutral policies on agriculture is stronger than that of the sector-specific policies". Both statements will be reflected by our analysis in that we will look into the economy-wide effects set in motion by *sector-specific agricultural policies* and, hence, their distortion potential, and secondly by analyzing the effects of *sector-neutral policies* or economic developments on agriculture.

The question of interest for our analysis is to what extent the general factors listed by Mundlak have also effected economic development of agriculture in the transition period and what *explanatory power* they have with respect to agricultural development in the context of transition? In the light of what has been said about the transition process earlier in this section, it is also interesting to ask how the various reforms which affected the economic environment for agriculture in transition might have complemented and/or reinforced each other. A conceptual framework for addressing these questions is provided by HÄGER, KIRSCHKE and NOLEPPA (2000) who discuss the market *forces affecting the supply of agricultural products in the transition process in a partial equilibrium framework* (see Figure 2-8).

Under initial conditions of the centrally planned economy, quantity q_1 is produced. The underlying assumption on the behavior of producers could have been cost coverage, input maximization or other principles that were typical for the way firms acted under the conditions of the socialist economic system. In the transition process and, hence, because of the abolishment of the central plan and the *development of a competitive system*, it is assumed that the producers changed their behavior in such a way that they now produce on the supply curve S_1. This *behavioral effect* is likely to result in a reduction of output, as firms have to adjust in various ways. For instance, the usage of variable inputs, which previously was determined simply by the size of the production unit or on the basis of historical information, is subject to marginal cost considerations when under the rule of decentral market coordination.

A second major effect is a *policy and price effect*: the high level of price support granted to agriculture in many countries at the end of the socialist period was abolished, which resulted in a strong decline in agricultural output prices; at the same time, prices for inputs increased, which was referred to as the *widening of price scissors*. While the reduction of price support is simply an *alteration of the price level* – for most agricultural products, real producer prices actually declined significantly in the course of transition – the second source of the price effect is an *alteration in the terms of trade.* In fact, price changes are frequently induced by policy changes but can also have different causes. While many agricultural policies have a rather complex institutional design, we will concentrate on the effects these policies have on prices and analyze the consecutive repercussions of such price changes in the general equilibrium.

Furthermore, a *technology effect* is to be expected: in the long run, the determination of relative prices by decentral market coordination should enhance the adoption of new technology, which would be associated with an outward shift of the supply curve.

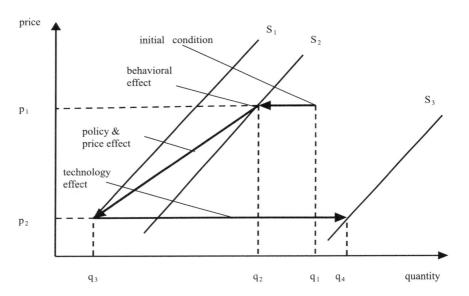

Figure 2-8   Effects of market forces on agricultural supply in the transition process

Source: Adapted from HÄGER, KIRSCHKE and NOLEPPA (2000).

This sequence of price changes results in a completely new formation of supply, indicated in the graph by the new supply curve S_3, and the new quantity q_4 which would be supplied after the shift from plan to market were completed at the new price p_1. This observation should be interpreted cautiously for at least two reasons: first, it is obvious that the graphical representation of the adaptation path is only one of many possibilities. For instance, while HÄGER, KIRSCHKE and NOLEPPA (2000: 254) assumed a positive terms-of-trade effect, the graphical representation in Figure 2-8 assumes a negative TOT effect, as it has been observed in the Russian context. Second, while the graphical representation shows a sequence of events, the various factors at work might actually overlap, reinforce, or even offset each other. Hence, the final result of these forces on agricultural development should also be a matter of a quantitative analysis.

The effects described in this partial equilibrium framework could be decomposed further. This is particularly so if the analysis is extended to a general equilibrium framework. Then the 'price and policy effect', for instance, could refer to alterations of various prices which in fact have occurred simultaneously in the course of transition and which have affected the development of agriculture. In the following section, we will therefore discuss how the above-mentioned effects can be analyzed within a general equilibrium model.

Analysis of transition related agricultural sector issues in a general equilibrium framework

Generally, it will be difficult to model the above-mentioned *behavioral effect* in any comparative static general equilibrium framework, particularly as this effect is more dynamic by definition than, for instance, policy changes. The analysis of behavioral effects will be constrained by the fact that we will not engage in a comparison of economic situations 'before and after transition' but instead in a comparison of two situations within the transition period. The reference year for which the model will be specified is 1994. Hence, it is assumed that the switch from pre-transition behavior to post-transition behavior has by and large already materialized. For reasons of simplicity, it is assumed that any other changes in the behavior of economic agents are exogenous to the model. However, because of the various structural changes in Russia's transition process, the stylized representation of agriculture in one sector as represented in Figure 2-8 would not be satisfying. Therefore, in chapter four we will describe how we have disaggregated Russia's agro-food sector according to various farm types (see chapter 4). This feature will be important in that, for instance, agricultural sector policies in Russia's transition process have mostly been directed at the former collective farm sector only, while the small-scale producers have been widely insulated.

In the framework, we propose the above-mentioned *price and policy effects* can then be taken into account directly in that various prices can be changed exogenously. Particularly relevant will then be the identification of differences between large-scale and small-scale agriculture. Another advantage of the general equilibrium approach is the potential to identify various sources of price changes. On the one hand, the price effects can be due to changes in internal and external terms of trade. On the other hand, effects of policy changes are often simply the results of a wedge driven between producer and consumer prices or between domestic and foreign prices in the base period. These wedges can be altered exogenously and therefore are prone to policy simulations. Depending on the details with which these policies are modeled, these wedges can have rather different effects.

Agricultural support policies are again a good example. In the context of the WTO, typical agricultural sector policies such as import tariffs and export subsidies are grouped into the so-called 'amber box'. They drive a price wedge between domestic and foreign prices. In contrast, policies which are not, or at least not directly, trade distorting can be classified either in the 'blue box' or in the 'green box'. Policies for direct income, as granted to farmers in the European Union, are grouped in the 'blue box', while policies which generally enhance the market environment for agriculture, such as agricultural research, would be 'green box' measures. While the common objective of all three policies could be to provide support to agricultural producers, the economy-wide effects are likely to be rather different. Therefore it should be possible to identify some of these differences within a general equilibrium framework; the structural specification of the GE model should alter some of these responses.

Consider the case of an increase in *sector support granted to agricultural producers with an import tariff*. In a general equilibrium framework, this should set various forces into motion: first, the increase in import tariffs would induce a terms-of-trade effect as described in the previous section. Domestic consumers would substitute imported commodities with domestically produced goods, depending on the elasticity of substitution. Second, because of higher domestic prices for the respective commodity. domestic producers will increase output depending on the degree to which additional production resources can be mobilized. In a relatively rigid model economy. backward and forward linkages will become particularly important. Third, in an import situation, the government should receive additional revenues. If government expenditure shares are fixed, those sectors which have already received the highest relative support in the base period would win. Fourth, the balance of trade has to be maintained, which means the real exchange rate or the capital account would need to adjust depending on the choice of foreign trade closure. This could induce 'second-round effects' in that any change in the exchange rate would imply relative price changes for all traded commodities. In contrast, effects on any non-tradable sectors in the economy would only be significant if aggregate demand changed in favor of the non-traded good. This effect would depend on the initial demand shares and again on the elasticity of substitution.

A third potential instrument would be *direct income support* and hence, a blue box measure. Support to agricultural producers would be financed exclusively by the tax payer and not by the consumer, which should have different implications for the economy as a whole. Most importantly, the deadweight losses which would be associated with the introduction of such a policy should be marginal because they should be the result of direct distortions in the economy-wide system of relative prices.

A *fourth policy instrument* with which support to agriculture could be granted and which therefore is also under discussion in the context of the WTO involves any kind of *measures which improve the market environment for agriculture* without directly providing support to agricultural producers. The best case in point is support for *agricultural research* and/or education. Normally, the returns on investments, for example into agricultural research, materialize only with considerable time lags. We therefore will not attempt to replicate the whole process of producing tangible agricultural research results within our model economy, but instead assume an exogenously determined effect on the productivity of a given sector and then look into the economy-wide effects of these productivity changes.

Technically, the effects of *green box policies* can be modeled in much the same way as proposed by HÄGER, KIRSCHKE and NOLEPPA (2000: 255) – by using a shift parameter in the production function to assess *technology effects*. In essence, a positive (negative) technology effect would be expected to yield higher (lower) output with a given set of inputs and hence be associated with an outward (inward) shift of the supply curve. Such effects can be related be partial productivity changes if the productivity change relates to a single production factor such as labor only. Total Factor Productivity (TFP) instead would look at output growth that is associated with changes in the amount of inputs in relation to all production factors. Quite obviously the institutional environment in the transition process is subject to many changes which not only alternate the costs of inputs but also technology and total factor productivity. Such changes would affect marginal costs and, hence, would have impacts on the supply curve of any commodity which is specified in the general equilibrium framework. This will enable us not only to look into the effects of shifts of the agricultural supply curve but also to analyze the implications of such shifts in other sectors. For instance, in a spacious country such as Russia, the transportation infrastructure is likely to be very important for the agro-food sector. This is so because food commodities are mostly bulk commodities and are characterized by low value density. Using the effects of shifts of the supply curve in the transportation sectors, we will analyze the effects of such changes on agriculture.

This discussion should indicate that there are more similarities than differences in the factors that drive agricultural development in general and in the transition process in particular. While the causes for these effects, particularly institutional reforms, might be rather different, the effects are not. A general equilibrium framework should therefore provide an adequate framework to analyze the effects of transition related factors.

Conclusions

This chapter set out to highlight some of the specific features of the transition process. A cross-country comparison indicated that the transition period has yielded a strong decline in output in terms of GDP and GAO. In Russia, the output decline of the economy as a whole and of agriculture has been particularly blunt. The specific path of transition has shaped the country's economy in a specific way: on the one hand, markets were liberalized quickly; on the other hand, decentralization occurred parallel to liberalization and became a major origin of structural rigidities in Russia's economy. These rigidities might actually explain why Russian markets have not always reacted to policy changes or exogenous shocks in the past decade as expected. Based on these observations, we argued that a *general equilibrium model that is based on 'neoclassical structuralism'* instead of on pure neoclassical economic theory will be more suitable for representing a transition economy such as that of the Russian Federation. We showed that such a model can theoretically provide meaningful insight into the economy-wide effects of agricultural sector development and vice versa. We also reviewed a conceptual framework for classifying the economic forces shaping agricultural development in transition. This discussion yielded two arguments: *first, the factors which drive agricultural development in the transition process might not be so different from the factors which have been shown to be important for agricultural development in the process of economic growth.* In fact, this might be another argument for the perception that *the difference between transition economics and development economics is not so great*. And, *therefore a theory of transition need not be a completely new theoretical body* but should draw heavily on complementary mainstream economic theories such as neo-classics, new institutional economics, and public choice (WEHRHEIM and VON BRAUN 2000: 527). Second, *the proposed general equilibrium framework will be suitable to look into many transition-related economic changes*, particularly those which have resulted in changes in relative prices that drive the decision of microeconomic agents. As we are also interested in quantitative insight into these issues, we will describe in the following chapters how the theoretical and empirical version of the model can be compiled in *an attempt to draw an economic map of the Russian economy in transition*, despite various remaining obstacles.

Chapter 3

A Computable General Equilibrium Model of Russia's Economy

This chapter discusses both the procedure of developing a general equilibrium model for Russia and its theoretical structure. The purpose of developing such a model is to obtain a tool with which the effects of exogenous shocks on the Russian economy and/or the country's agro-food sector can be analyzed quantitatively. Whenever conclusions based on the model's 'output' are made, it has to be clear that these always depend on the model's 'inputs'! This fact necessitates a high degree of transparency in order to enable the model user to trace the economic causalities which drive the results in the model world. There are many links within the model, which often makes it impossible or at least tedious to trace the respective effects using only one's own mental arithmetic. We therefore also expect the model to give counter-intuitive results! A counter-intuitive result can highlight an economic causality or interaction that previously had been undervalued or even neglected. As the model is based on straightforward equilibrium theory, such surprises should be empirical instead of theoretical ones (DEVARAJAN, LEWIS and ROBINSON 1994).

The origin of computable general equilibrium models

The roots of general equilibrium thinking in economic history can be traced back at least to the French physiocrats. Quesnay's concept of the circular flow of resources in an economy, as presented in his *Tableau Economique*, already identifies economic links which later became integral parts of general equilibrium theory. Adam Smith's metaphor of the 'invisible hand' was and still is another important concept which hints at the general equilibrium considerations intrinsic to classical economic thought.

The neoclassical general equilibrium concept, however, was developed over a long period, beginning with the work by Walras and culminating in the seminal studies of Arrow and Debreu, of whom the latter, in 1983, received the Nobel Prize for his contributions to this concept. Indeed, Léon Walras (1834–1910) was the first economist who succeeded in developing a formal framework for the market equilibrium. In the first and second volume of his chief work, WALRAS (1874 cf. in 1988) developed his theory of a market equilibrium. In doing so, he had to solve two problems: firstly, to prove that an equilibrium between supply and demand is possible and, secondly, that it is stable. To solve the first problem, Walras showed that a social optimum in a market will be reached when the relation of marginal utilities of all demanded goods is equal to the relation of all prices. Without formally proving the latter proposition, Walras relied on prices as the major variable for achieving an equilibrium and introduced another grand metaphor of economic theory, the 'auctioneer', who helps in identifying the equilibrium price step by step. The well-known central conclusion of this *first Walrasian law* is that if a market is free and, hence, competitive, it will find the social optimum.[1] It indicates that the respective allocation of such an equilibrium is pareto-optimum and that all benefits of exchange have been exhausted. To address the second problem, Walras represented demand and supply in a functional form. He showed that if the number of equations in the model equals the number of variables, the model can have one solution only, which means that in equilibrium, and given that the assumption of convex preferences holds, only *one* price (or *one* price vector in a multi-market framework) will yield the optimum solution. This again is known as the *second Walrasian law*. In fact, these central features of general equilibrium theory indicate why free and competitive markets are expected to be superior to centrally planned ones: not only because they yield a social optimum but also because the use of the price mechanisms seems to be the most efficient way in compiling the information required to find it. If all parameters in the equations of a Walras' model were known, and the equilibrium were static, the central planner would be able to reach the same optimum. Indeed, the *price mechanism* which replaced the social, or better central, planner in the transition period must also be understood as *the most important source of market information*. An important implication of these considerations is that any policy should avoid altering the price mechanism. Pure neoclassical theory, however, would not have any argument against a redistribution policy of the state which transfers purchasing power between various population segments for equity considerations (cf. VARIAN 1991: 480).[2]

Another important point is the fact that *general equilibrium models*, even though rooted in the same theoretical core concept mentioned above, are a *continuum* rather than a group of homogenous models. Indeed, the long history and the wide range of applications requires a distinction between different types of general equilibrium models. For a long time, purely *theoretical models* including only two or three sectors were dominant (Figure 3-1). They were extensions of the Walras model by looking at links between various product and/or factor markets and were helpful in analyzing the general economic causalities within a general equilibrium framework. Many of these theoretical general equilibrium models were purely neoclassical in the sense that prices varied and goods and production factors were fully mobile across markets. The next generation of these models were stylized representations of groups of countries, for example, developed countries. They included a few sectors and stylized numerical facts about a specific economic setting to base the analysis

---

[1] In the third part of his chief work, Walras also wrote about the conditions for asset distribution (land) which have to be fulfilled to reach the social optimum. However, this part of Walras' work was never accepted as an integral part of neoclassical theory, but instead has been neglected.

[2] Indeed, this theoretical argument is the basis for the critique of the agricultural subsidy systems in many countries. In the European Union (EU), for instance, for a long time, subsidies were granted to farmers via price support. These distortions of the price mechanism led to tremendous misallocations before the 1990s agricultural policy reforms in the EU began to depart from the old system of price support and slowly switch towards direct income support via direct payments.

on empirical observations. Such *stylized models* were built to better understand problems too difficult to be solved by mental arithmetic alone, or problems that have ambiguous answers depending on the scope of model parameters. Furthermore, stylized models not only allow the identification of the direction of economic causalities, but also the quantification of the size of various effects. The third group of general equilibrium models which we refer to in this study are further extensions of stylized models and are termed *applied computable general equilibrium models (ACGE)*. Such ACGE or CGE models normally have a much more detailed data base which best captures the specific situation of a particular economy.[3] While stylized – and even more so applied – models are usually larger and more realistic than purely theoretical models, they stay close to their underlying analytic model ancestors (DEVARAJAN, LEWIS and ROBINSON 1994). On top of the progress made in developing the underlying general equilibrium theory, the emergence of applied models has certainly been accelerated by the greatly enhanced possibilities to process large amounts of data using computers (see chapter 4).

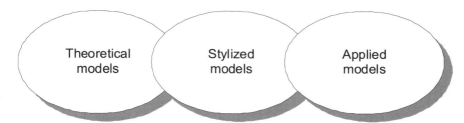

Figure 3-1   The continuum of general equilibrium models

One theoretical model which has been used as a reference for many stylized and applied models is the so-called *Australian model*, which is also one of the first examples of an applied CGE model (Salter 1959; SWAN 1960). The Australian model was innovative in distinguishing between *tradables* (including exports and imports) and *non-tradables*. Whenever the good from one sector was traded, its price was determined by world market prices. Even though the share of purely non-traded commodities is decreasing in the course of economic development, and as globalization encompasses an increasing number of commodity markets, this distinction will be useful for the description of some characteristics in the Russian economy.

Another important 'blueprint' for many of today's CGE models is the above-mentioned *1-2-3 model* which is discussed in detail in DEVARAJAN and LEWIS (1990) or DEVARAJAN, LEWIS and ROBINSON (1994).[4] In the 1-2-3 model, all domestically

---

[3]   Following a common convention, whenever we use the term 'CGE model' in the following we are referring to the applied CGE model approach unless otherwise stated.

[4]   In the Appendix the complete set of equations of the '1-2-3 model' is presented.

produced goods that are not exported are effectively treated as non-tradables or *semi-tradables* which are imperfect substitutes to one another both on the production and on the demand side. For given substitution and transformation elasticities, DE MELO AND ROBINSON (1989) showed that the domestic price in a sector is more closely linked to the respective world price the greater its trade share. In effect, the transmission elasticity between domestic prices and world market prices depends on the trade share.

The extension of these applied models went along with the incorporation of *structuralist features*. Even though such modifications are most frequently applied to developing countries, rigidities which exist in industrialized countries such as in labor markets could therefore also be appropriate components of CGEs for the latter. The respective extension of the first generation of neoclassical applied general equilibrium models was performed by incorporating *structural rigidities* in the economies of developing countries with which the functioning of the price mechanism was restricted. Even though there are many distinct differences between developing countries and transition economies, the approach to modeling structural rigidities will also be useful in characterizing the Russian economy.[5] These approaches can be useful templates for modeling economies in transition, revealing in many instances features similar to those in developing countries. In both types of economies, markets are characterized by distortions impinging on the price mechanism, which means that prices do not always adjust to an equilibrium freely and instantaneously.

Review of existing models for transition economies

When transition began, a new class of CGE models started to emerge. Most of these models built on the tradition of CGEs for developing countries because the economic distortions were often of a similar nature (cf. JÜTTING and NOLEPPA 1999). For instance, many developing countries that were confronted with severe macroeconomic imbalances in the 1980s had to implement structural adjustment measures. The most prominent features were exchange rate realignments and reductions in government budget deficits. CGEs became an important tool for analyzing the repercussions of the mandatory reforms to correct the respective imbalances in developing countries. Furthermore, the objective of using such models for policy analysis in developing countries was similar to the objectives of using CGEs for transition economies: while recognizing the limitations of CGEs, they are the most suitable device for tracing allocative and distributive effects of macroeconomic shocks on the behavior of various agents in the economy (e.g. ROBINSON 1991; SADOULET and DE JANVRY 1995; SHOVEN and WHALLEY 1992).

---

[5] Differences between developing countries and transition economies refer to various stylized facts. The most pervasive differences are: First, on the eve of transition, the former centrally planned economies mostly revealed an industrial and production oriented structure instead of being an economy with a dominant agricultural sector; second, while the state was omnipresent, an institutional framework based on market principles was almost non-existent; third, the human capital base in transition economies was in a comparatively advanced state due to relatively broad provision of primary and secondary education.

Since the late 1990s a whole range of *CGE models* has been *developed for transition economies*. However, the geographical west-east divide in the level of economic development of transition economies in Eurasia (cf. WEHRHEIM 2000b) and its positive correlation with the quality of statistical information seem to be responsible for the fact that the further east one goes in the region, the more difficult it becomes to model the respective economies. This is also reflected in the number of CGE models which have been developed for the respective countries in the region in the more recent past.

A relatively large number of CGE models has been put forward for the CEEC countries, and, particularly, for the four Visegrad countries. Even though these models are often not designed to exclusively address agro-food sector issues, they refer to the sector for two reasons: first, it still constitutes an important sector in most transition economies; second, agricultural policy issues are of particular relevance in the context of CEEC's integration into the EU.

Most of the CGE models developed for transition economies are one-country models (TESCHE 1994; BANSE 1997; SCANDIZZIO 2000). However, especially when addressing the EU integration aspects, a multi-country framework also seems suitable. Such a model has been proposed by WEYERBROCK (1998), involving four regions: the EU, the CEEC, the US and the Rest of the World. SWAMINATHAN, BROCKMEIER, and HERTEL (1998) used the modeling framework of the Global Trade Analysis Project (GTAP) to analyze the economic impact of CEECs acceding to the EU. Many of these models suffer from the essential trade-off between partial and general equilibrium models with respect to sector-specific details. In order to better address the economic implications of, for instance, liberalization strategies in the agro-food sector of transition economies, partial equilibrium models have been proposed (FROHBERG et al. 1998; FOCK et al. 2000). In an attempt to reconcile the advantages of partial and general equilibrium models, BANSE and MÜNCH (1998) linked respective modules for the Hungarian economy and included internal feedback loops.

What is lacking to a great extent in most of these *CGE models* is a *disaggregation of primary agricultural production* that would be based on the differences between various farm types. Particularly in the CIS countries, a *dual agricultural production structure* has emerged in the course of transition in that the former collective farms and the newly emerged private subsidiary plots coexist today (see chapter 1). Nevertheless, in most CGE models, primary agricultural production is either not disaggregated at all or according to products only.

Procedure in constructing a CGE model for transition economies

The discussion in the last sections indicates that there are various options once the decision has been made to use a CGE to analyze the specific economic problem at hand. A distinction can be made between at least three major phases of the model building exercise, each consisting of various major steps (Figure 3-2).

Phase 1 is the initial phase in which major decisions based on the knowledge of the country and the economic problems to be analyzed have to be made as regards the purpose and objectives of the modeling exercise. Here, the modeler should define the *objectives and purposes* of the modeling exercise (Step 1). The model choice and the economic and empirical details to be included in the model in later steps crucially depend on the nature of issues and the sectors on which interest is focussed. In the case of transition economies, this is particularly relevant as a decision has to be made as to which model will be capable of revealing some of the features that are important for these economies. For instance, it has to be decided up-front if the model will focus on distributional effects of specific policies or if, instead, allocative repercussions are to feature more prominently. This would have implications for the *type of institutions* represented in the model (Step 2). A model which intends to look at distributional issues should differentiate between a number of households, while a model whose prime objective is to look into allocative effects of various policies would not need to have all that many details on the household sector but should put more weight on the disaggregation of the economy.[6] Another possibility would be to analyze the implications of various foreign trade-related policies on the welfare of various trading blocs. In fact, this type of multi-country CGE model, some of which are often combined to form world trade models, has received a great deal of attention. They feature prominently, for instance, in the above-mentioned GTAP-model (HERTEL 1997).[7] Therefore, a decision also has to be made at this stage as to the kind of policy experiments that are of interest. Obviously, trade-related experiments would necessitate a completely different model structure than experiments focusing exclusively on domestic policies.

Phase 1	**Step 1:**	Definition of model purpose and objectives
	Step 2:	Specification of institutional actors in the model
Phase 2	**Step 3:**	Formulation of theoretical structure of the model
	Step 4:	Compilation of a data base for a single year and testing for consistency
Phase 3	**Step 5:**	Simulation of policy experiments and exogenous shocks
	Step 6:	Policy appraisal based on pair-wise comparison between counterfactual and reference situation

Figure 3-2   Steps in constructing an applied general equilibrium model

Source: Adapted from Wiebelt (1996: 72).

---

[6] It is common practice to use the phrase *institutions* when referring to households, firms, and other economic groups of agents represented in CGE models. However, this neglects the broader definition of the term as known today from *institutional economics*. Here, the term institutions, refers to the rules, norms, and values which define the way economic agents communicate with each other in *organizations*. Hence, strictly speaking one should refer to households and firms etc. as economic organizations.

[7] The GTAP offers the great advantage of being accessible via a public domain to which any modeler can have access. It is particularly convenient for cases in which the data base for the individual countries of interest are already available. This is, however, not yet the case for the Russian Federation nor for any other countries of the FSU.

In *Phase 2*, the modeler should deal with the actual design and construction of the model economy. This is normally the most demanding and central phase in the model building process. Based on the defined policy issues and the chosen level of disaggregation, the theoretical model has to be formulated in a functional form which is both theoretically consistent and reveals reality in such a way that the model can be validated in later steps. As indicated in chapter 3, the economic starting point of CGE models is Walras' neoclassical world. Mathematically, standard CGE models consist of a set of non-linear and simultaneous equations (LÖFGREN 2000). Once this functional representation of the complete theoretical model has been specified, it has to be translated into a programming language as the basis for solving the model. Hence, *the* general equilibrium has to be found by solving the formulated algorithm in iterative steps. For the implementation of the Russia model, we have used the GAMS software (cf. BROOKE, KENDRICK and MEERAUS 1988; RUTHERFORD 1998).

Closely linked to Step 3 is Step 4, the collection of data and the compilation of a consistent data set. Due to relatively poor data availability, this step indeed often proves to be a daunting task, particularly for transition economies. Generally speaking, the data consists of three major elements. First, macroeconomic data has to be compiled. Second, the input-output matrix for the sectors defined in Step 2 as well as the sectoral break-up of macro totals has to be carried out. Both elements have to be combined and must yield a consistent data set, the so-called micro social accounting matrix, which should represent all economic transactions in an economy in a given year in such a way that theoretically established equilibrium conditions are met. Third, the parameters, i.e. mainly the various elasticities, have to be either estimated, calibrated and/or synthesized from the literature. Once a consistent data base has been compiled, it also has to be transformed into a GAMS file and can then be linked to the theoretical part of the model.[8]

Finally, in *Phase 3*, the theoretically and empirically consistent model can be used for its true purpose: carrying out simulations and policy appraisals with the model. Relating to phase one, the simulation exercises should now be specified. Step 5 should also include sensitivity tests which could consist of, for instance, an alteration to trade parameters. Various options are given for simulating policy scenarios: on the one hand, single variables can be altered exogenously, e.g. tariff rates, the exchange rate etc.; on the other hand, the closure rules of the model can be altered. A third option is any combination of the alternatives mentioned. The results of the simulations are compared against the base run and have to be interpreted by explaining the forces which drive the adjustment to exogenous shocks within the model. It is quite obvious that there might also be feedback loops between the various steps involved in the model-building process. The possibility to carry out positive and normative simulations with a base run version of the CGE is another attractive feature of the approach.

---

[8] The detailed steps in compiling the data base for the Russia model will be elaborated in chapter 4.

Objectives of the model and institutional design

The focus of this model for the Russian economy is on agriculture and its role in the transition process. The objective is to explain the sector's restructuring *and* to shed some light on foreign trade-related options. Generally, we will be interested more in allocative issues in this study and not focus on distributional matters (see also chapter 1). Therefore, only one household will be specified while we will pay more attention to the disaggregation of the economy into various sub-sectors. Particular weight will be given to primary agricultural production, which experienced significant restructuring during the transition period. Furthermore, because we intend to look at foreign trade-related matters, the relevant institutional features should be represented in the model. On the one hand, relations with the rest of the world should reveal trade flows, exchange rate effects, and effects on foreign capital in- or outflow. Imperfect substitutability between domestic goods and traded goods should also be presented by the model as this constitutes an important feature of the Russian economy. On the other hand, trade instruments such as import tariffs and export taxes should be modeled to be able to analyze various trade strategies currently debated in the Russian Federation.

Theoretical structure of the model for Russia

Overview of the model structure. The theoretical structure of the multisector general equilibrium model for Russia is an extended version of the above-mentioned simple 1-2-3 model. It builds on the tradition of the first applied macro-structuralist CGE models (TAYLOR 1990). These models have been extended to micro-structuralist models which give more attention not only to the macroeconomic features but also to the microeconomic behavior of various economic agents (DERVIS, DE MELO and ROBINSON 1982). One of the first country applications of the latter type of model was developed by ADELMAN and ROBINSON (1979) for South Korea; this later served as a 'blueprint' for many other country studies for quantitative analysis of the effects of economy-wide reforms on agriculture.

The Russia model builds on the tradition of these models. It depicts the *circular flow of economic resources* in the economy by representing the most important economic actors and the economic links between them in a stylized form. It therefore provides a snap-shot of the Russian economy in a given period and, hence, can be used in a *comparative static* manner. *Structural rigidities* are taken into account to reflect the limited mobility of production factors (see chapter 2). The model represents *four central economic agents* in the economy, whose behavior has been aggregated at differing degrees: enterprises, households, the government, and the rest of the world. It is the circular flow of economic resources between these institutions that is captured with the model. Obviously, the links between these economic agents are more complicated than indicated in Figure 3-3. Indeed, additional institutions are needed to model the circular flow in a structured manner. One example of such an additional institution is a stylized 'capital account', which basically represents a large big nationwide bank through which all savings-investment transactions are channeled. The capital account balance, the government budget surplus or deficit, as well as private savings and investment transactions are summarized in this account. Furthermore, the stylized representation of links between the economic institutions in the economy conceals far more complicated interactions between any pair-wise combination of economic agents. For instance, firms are likely to pay various forms of direct and indirect taxes to the state, while they may at the same time receive not only payments for goods and services sold to the government but also subsidies. As these flows between the government and the firms are likely to differ depending on the sector in which firms operate, the firm sector will be disaggregated to the greatest extent.

Hence, the production sector is split into several different sectors, revealing the sectoral composition of production in the Russian economy. Each sector produces one good which is representative for the sector but not necessarily homogenous. Instead, each sector can produce different goods for the domestic market and for foreign markets. This implies that sectors in the Russian economy use different marketing chains when delivering goods either to the domestic or to the foreign market. Producers in some sectors sell to both domestic and foreign markets depending on the relative prices in both market segments.

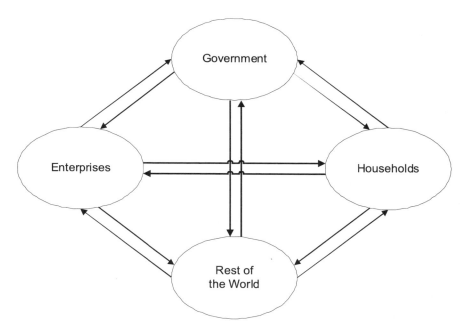

Figure 3-3   The institutional players in the Russia model

Furthermore, an important component of final demand is modeled by representing the behavior of households. Depending on their resources, households receive capital or labor income for the factors they employ in the firm sector. This income is used for consumption expenditures, for payment of taxes to the state and for savings. Consumption of the individual goods hinges on the nominal income and on prices. Private households differentiate between domestically produced and imported goods which are interchangeable to some degree.

Table 3-1 summarizes the components of supply and demand of goods in the model. All sectors in the economy together produce domestic supply which is allocated to the domestic market (domestic supply) but also to foreign markets (export supply). The other side of the model coin comprises the demand components. On the one hand, (domestic) demand consists of demand for intermediates, investments, and stocks by the production sectors. On the other hand, the government and private households demand the goods supplied by the domestic production sectors. At the same time, domestic demand can also be satisfied by buying imported goods supplied by the rest of the world to domestic markets. Export demand is another component which directly matches domestic export supply and is restricted to tradables.

Table 3-1   Components of supply and demand in the model

Domestic Supply Components	Domestic Demand Components	International Supply Component
Domestic supply	Intermediate demand Investment demand Demand for Stocks Household demand Government demand	Import supply
Export supply	Export demand	

Source: DERVIS et al. (1982).

Hence, the rest of the world is also an active institution in the model on both the supply and the demand side by buying tradables from and selling them to Russia. Additionally, the newly gained 'openness' of the Russian economy is reflected by the capital account enabling various types of transactions between Russia and the rest of the world. These features allow us to specify, for instance, capital flight, which has become an important characteristic of Russia's economy in the 1990s, within the model. Total investments are determined endogenously via the total sum of savings, including the deposits of foreign investors. Changes in stocks depend linearly on production. Real demand of the state is determined exogenously. Adding demand over all of these components yields total demand. In a second step, total demand is split into demand for the domestic and the foreign good. This split depends on the distribution of respective quantities in the initial situation, the relative prices between domestic and foreign commodities, and a parameter defining the substitutability, which is equal across all five domestic demand components.

Choice of functional forms

When designing a CGE model, one of the most important decisions is the choice of functional forms. Obviously there are various options in the way of functional forms for the representation of the behavior of various actors. Due to the wide selection of functional forms that are possible, the decision can be based on various criteria such as:

- the flexibility of respective functions,
- the number of parameters that have to be estimated,
- the possibilities to interpret the results obtained with specific functions,
- the possibilities to estimate the complete function.

The flexibility of the functional forms refers to the degree of substitutability between the various explanatory variables. On the production side, a Cobb-Douglas function, for instance, has a much lower degree of substitutability than, for instance, a Generalized McFadden function (for a model using this function, see e.g. Fock et al. 2000). However, the number of parameters that have to be estimated increases with the flexibility of the function. If such parameters are not available from econometric estimates, they have to be calibrated. In order to get meaningful parameters with such a calibration procedure, start-values have to be used, and these either have to be compiled synthetically from other studies on comparative countries or they are simply 'sophisticated guesses'. In the context of transition countries, and particularly Russia, empirical estimates of parameters are almost non-existent. Generally, the choice of functional forms should therefore also be a function of the data availability: the less data is available, the simpler the functional forms should be. As a consequence of the data available for Russia, we will use rather inflexible functional forms that necessitate the predetermination of only a minimum number of parameters. Additionally, results obtained with such simple functional forms can be interpreted much more easily, as the causality within the model can be traced more directly. A final argument for the choice of relatively inflexible functional forms for the Russia model is the fact that, even though sophisticated functions may be theoretically more convincing, their explanatory power for transition economies has not always turned out to be greater.

Figure 3-4 summarizes the *functional forms used in modeling supply and demand*. On the supply side, X_n products are produced by the various sectors in the model. The production of each good is allocated using a Constant Elasticity of Transformation (CET) function between exports (E_i) and supply for the domestic market (D_i). The good supplied by each sector is produced using primary factors and intermediate inputs, the split between the two being defined a priori with a Leontief production function, excluding the possibility of substitution between primary factors and intermediates when the model is exposed to shocks. The intermediates in each sector (V_i) are composed of domestic and imported intermediates based on a Constant Elasticity of Substitution (CES) function with a limited substitutability. Demand for primary production factors is derived by minimizing costs subject to a Cobb-Douglas technique. Producers in each sector decide upon factor use according to the prevalent factor prices, output prices and production taxes.

Export demand is price elastic, which implies that Russia can have an impact on world market prices. Depending on the sector-specific situation, the country can be modeled as a *large country*, which will be particularly relevant for raw material exports. Import supply is completely elastic, assuming that Russia's imports in the transition period comprise only minor shares of total world trade in the respective sectors. Taking this assumption for granted, it would be unrealistic to expect that changes in Russia's import demand would cause any changes in world market prices. The exchange rate is modeled to indicate the transmission between domestic and international prices.

Household demand depends on both consumer expenditures and commodity prices. Consumers maximize utility subject to a budget constraint and fixed budget expenditure shares for the composite good produced by n sectors in the model economy. The composite good consumed from each sector is represented by nested CES functions which allow substitution between the domestic (D_i) and imported good (M_i).

The degree of substitutability or complementarity of domestic and foreign goods is one distinct feature of micro-structuralist models as opposed to neo-classical models. While the original neoclassical model assumes perfect substitutability, micro-structuralist models assume limited substitutability between domestic and foreign products. While the former assumes that domestic product prices are totally dependent on world market prices, the latter assumes that domestic prices are determined by world market prices, the exchange rate and possibly by trade restrictions.

In the Russia model, both of these two extremes are possible, as are solutions ranging in between these two poles, depending on the characteristics of the respective sector. As a result of this differentiation between the domestic goods as well as imported and exported goods in each sector, the model which has n sectors can represent a maximum of $3n$ goods. Indeed, the assumption of limited substitutability between domestic and foreign goods seems to be of relevance for an economy in transition, and particularly when analyzing Russia's agro-food sector in the transition period. Strong preferences of Russian consumers for branded food products imported from western countries, for instance, were one reason for the country's increasing deficit in agricultural and food trade in the 1990s.

Equations of the model

In general, the behavioral equations of the model are similar to those of many other single-country and multi-country models in that they use functional forms such as Cobb-Douglas, CES and CET to specify the behavior of the agents in the economy (cf. LEWIS, ROBINSON, and THIERFELDER 1999: 14; WOLF 1996). The equations of the Russia model will be presented and discussed by grouping them around the major economic activities that reveal the circular flow of income in the economy. First, the functional representation of the production side will be discussed. Second, the distribution of income generated in the production process will be described, leading directly to the description of household behavior. Since the behavior of both firms and households are also linked to the rest of the world, these links are discussed next. What follows is the specification of the behavior of the third important agent in

Supply side

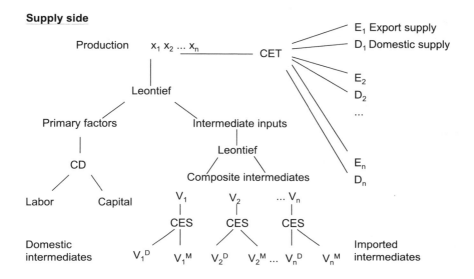

Demand side

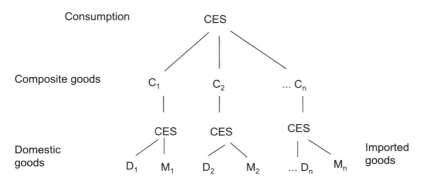

Figure 3-4   Overview of model structure

the model, the government. Furthermore, the savings behavior of the agents in the model will be defined. Then, some additional macroeconomic equations for gross domestic product are presented to balance the model and define the *numeraire* of the model. Finally, the closure rules of the model will be elaborated.

The following rules on notation will be applied for the definition of variables (see also Tables A3-3 and A3-4): endogenous variables are represented with capital letters (e.g.: P), while exogenous variables are typified as capital letters with a bar (e.g.: $\bar{P}$). Policy variables such as taxes or tariffs are indicated using small letters (e.g.: t). Furthermore, a distinction is made between real and nominal variables (e.g.: $\tilde{S}$). A further convention is to use small letters to define, for instance, parameters (e.g.: a_i) specifying a specific feature in the respective sector (e.g. partial elasticities of production factors). Small letters of the Greek alphabet (e.g.: δ_i) represent the various shares of n sectors of specific variables in the base run scenario. Additionally, super- and subscripts are used. Most superscripts refer to the regional classification of the variable. The superscript m, for instance, indicates that the relevant variable is defined for imported commodities (e.g.: P^e, i.e. price for exports; P^d, i.e. price for domestic goods). Furthermore, for the representation of world market prices in foreign currency the superscript $\$$ is used (e.g.: $P^{\$e}$, i.e. world market price of exports in foreign currency). The subscript i and j represent the 1 ... n economic sectors in the model. If both subscripts are used the i indicates that the respective variable represents the *originating* sector, while the subscript j hints at the *receiving* sector. The subscript f represents the different production factors (capital and labor) and p the private households in the model.

Functional representation of production, demand for primary factors and intermediates, and investment demand

One of the common features of CGE models is the assumption that firms maximize revenues and minimize costs and, hence, maximize profits. All sectors in the economy are characterized by constant economies of scale. Furthermore, we assume perfect competition between the enterprises in each sector. The assumption of profit maximizing enterprises is also used in this model, but at the same time several restrictions have to be taken into account and this is done in various steps. Moreover, we assume constant scale economies. Many of these assumptions have been said to be too unrealistic to represent transition economies properly, and they have been contested in other models for various transition economies (e.g. PAVEL 2000; NUPPENAU and WEHRHEIM 1999). However, there is no convincing empirical evidence in the literature that any other economic paradigm would have the same or a higher explanatory power with respect to the Russian Federation than the assumption of profit maximization and perfect competition. We will therefore rely on this functional representation, while at the same time acknowledging the possibility of different behavior in individual segments of the Russian economy and/ or regions of the country.

The production volume in each sector of the Russia model is determined in three steps. First, the sectoral production volume is modeled with fixed shares of a bundle of primary factors and intermediates, which precludes substitution between these two production factors. Substitution between various intermediates is also zero, as fixed input output coefficients (a_{ij}) are used according to a Leontief specification. Demand for the sector-specific intermediate good (V_i) is then calculated by totaling the respective input-output shares over all sectors, multiplied by the amount of output in the sector using the intermediate commodity:

$$V_i = \Sigma_j a_{ij} X_j \qquad\qquad\qquad (3\text{-}1)$$

As substitution between intermediates and the production factors F_{fi} (capital and labor) is precluded, the use of intermediates does not appear as a separate variable in the production function. Instead, on the second level, the production function specifies the sectoral demand for factors (capital or labor) in sector i and functions as an additional constraint on the profit maximizing behavior of the firms. Land is part of capital. With respect to human capital, it is also assumed that both population and behavior of producers is constant which implies that entrepreneurial capital is fixed as well. Substitution between labor and the factor capital is modeled with a Cobb-Douglas function using an elasticity of substitution of 1:

$$X_i = a_i{}^x \Pi_f F_{fi}{}^{\alpha_{if}} \tag{3-2}$$

The parameter α refers to the partial production elasticities of the various production factors employed in the different sectors and due to the adding up constraint must equal 1 for each sector. The parameter a_i is endogenously calibrated and indicates the efficiency of the production factors in the respective sector, and, hence TFP. This parameter will play an important role later on in the simulations as it can be used to simulate shifts of the production function (see chapter 5).

On the third level, the substitution between imported and domestically produced intermediates is modeled with a CES function. The elasticity of substitution used in the model is the same for all components of final demand. Hence, the degree of substitutability is the same, no matter whether final consumers, production branches or other domestic demand components use the respective commodity. Assuming that the firms minimize costs, demand for labor depends on the wage rate for labor employed (W_f). This is the nominal wage rate which is determined when the demand for labor and supply of labor are in equilibrium. The parameter ϕ_{if} is a shift parameter which indicates the difference between the average and the sector-specific marginal value product of labor. Hence, demand for labor is determined with the following equality, indicating that marginal value product of labor is equal to the factor price:

$$W_f \phi_{if} := \cdot P_i^v \cdot \alpha_{if} \cdot \frac{X_i}{F_{if}} \tag{3-3}$$

If the prices for the sector-specific value added (P^v), and the price for the wage rate (W_f) are given, labor demand is fully specified in the model. Therefore, factor prices are determined on factor markets and function as signals to the various production sectors employing the factors. If factor prices change because of an exogenous shock, the various sectors in the economy respectively adapt factor demand.

However, as discussed below, producers determine the total volume of production not only by supplying their products to the domestic market. Instead, they maximize their revenues from sales to the domestic *and* foreign markets. The respective production possibility frontier is modeled using the restriction of the following aggregation function, exhibiting a CET typology. In the following function, a_i and δ_i are efficiency and distribution parameters, respectively. The distribution parameter δ indicates the split of X_i into exports and supply to the domestic market in the base run scenario, while the transformation parameter is represented by ρ_i. The value of the elasticity of transformation indicates the degree of homogeneity of the commodity supplied to the foreign and domestic market:

$$X_i = a_i \left[\delta_i E^{\rho_i} + 1 - \delta_i D_i^{\rho_i} \right]^{1/\rho_i} \tag{3-4}$$

As producers distinguish between production for the domestic and the foreign market, the sectoral producer price (P_i^x) is a weighted average of the price in each of these market segments (domestic and foreign):

$$P_i^x = \frac{P_i^d \cdot D_i + P_i^e \cdot E_i}{X_i} \tag{3-5}$$

From the equations above, it is evident that the demand for production factors does not depend on the gross price in the sector but instead on the sectoral net or value added price (P_j^v) which therefore has to be defined first. In fact, this price is instrumental in identifying the relative changes in the profitability of primary production factors and will also be reported when the results of various policy experiments will be discussed in chapter 5. It is derived from the total output price (P_j^x) by deducting direct taxes (t_j^x) and the expenditures for intermediates:

$$P_j^v = P_j^x (1 - t_j^x) - \Sigma_i a_{ij} P_i^q \tag{3-6}$$

As indicated in Table 3-1, another important component of total domestic demand is demand for investment goods, which can be partitioned into two sub-components: first, demand for investment goods and, second, the change in stocks. The sectoral share for investment demand is assumed to be constant in the framework of this comparative static general equilibrium model. The demand for investment goods i in sector j is dependent on the pre-determined sector-specific coefficients (b_{ij}) and the gross investments in sector j:

$$I_i^A = \Sigma_j b_{ij} \cdot I_j^B \tag{3-7}$$

Another relevant price which has to be defined is that for the capital good. It is calculated as the purchase price of the composite consumption good (P_i^q) in sector j, weighted with the sector-specific share of the capital good (b_{ij}) which have to add up to one:

$$P_j^k = \Sigma_i b_{ij} P_i^q \ with \Sigma_i b_{ij} = 1 \tag{3-8}$$

Furthermore, it is assumed that stocks (I_i^w) are determined as fixed shares (β_i^t) of total sectoral production (X_i):

$$I_i^W = \beta_i^t X_i \tag{3-9}$$

Total gross nominal investments can then be calculated as the residual between total nominal investments and the sum of sectoral stock building. In equation 3-20 the latter is calculated by multiplying the sectoral output (X_i) by the sectoral share of stock building in sector i in the initial situation (β_i^I):

$$\tilde{I}^B = \tilde{I} - \Sigma_i \beta_i^I X_i \tag{3-10}$$

What remains, is the definition of real investments in each sector of the economy. It can be specified as the share of the sectoral investment goods used in sector j (β_j^B) times gross nominal investments ($\tilde{I}^B$) divided by the sectoral capital good price (P_j^k). Again, the sum of the sectoral investment shares have to add up to one:

$$I_j^B = \beta_j^B \tilde{I}^B / P_j^k \ with \Sigma_j \beta_j^B = 1 \tag{3-11}$$

Based on the above definitions, we can now model the profit maximizing behavior in each sector under the restrictions of the nested production function as outlined above. All the relevant components have been defined and the total firm profit can be calculated as total revenues ($P_j^x \cdot X_j$) minus indirect taxes ($1-t_j^x$), minus the costs of intermediates ($\Sigma_i a_{if} \cdot P_i^q$) and minus the costs for labor, which depend on the amount of labor employed in the respective sector (L_j) and on the respective wage rate (W):

$$\pi_j = \left[P_j^x \cdot (1 - t_j^x) - \Sigma_i a_{ij} \cdot P_i^q \right] \cdot X_j - \Sigma_f f \cdot W_f \cdot \phi_j \cdot L_j \tag{3-12}$$

This equation can be simplified by including the value added into the function, from which direct taxes and labor remuneration have to be deducted:

$$\pi_j = P_j^v \cdot X_j - \Sigma_z (1 + t_j^l) \cdot W \cdot \phi \cdot L_j \tag{3-13}$$

Allocation of income

In the next step, the income resulting from the production process has to be distributed among the various relevant actors. In general, three different institutional agents can be distinguished as recipients of this income: capital owners who receive profits, workers who receive salaries, and the government which receives shares of the income as tax revenues.

The remuneration of factors in each sector (Y_j) depends on the sector-specific nominal factor payments for each of the production factors (W_f) and the amount of the different production factors employed in the respective sector (F_{if}). Accordingly, total factor income (Y_f) can be determined by totaling these components over all production sectors:

$$\tilde{Y}_f = \Sigma_i W_f \cdot \phi_{if} \cdot F_{if} \tag{3-14}$$

From the total income which capital owners receive depreciation (A) has to be deducted to calculate the net income of capital owners:

$$\widetilde{Y}^{K}=\widetilde{Y}_{f=1}-\widetilde{A} \tag{3-15}$$

Total depreciation is defined as the sum of sectoral depreciation, which depends on sectorally different depreciation rates (d_i), the price of the capital good in each sector (P_i^k), and the demand for the capital good (F_i):

$$\widetilde{A}=\Sigma_i d_i \cdot P_i^k \cdot F_{i1} \tag{3-16}$$

However, in the model we are interested in the total income received from wage payments (Y^L) and therefore have to sum the respective income over all sectors:

$$\widetilde{Y}^{L}=\Sigma_{f\neq1}\widetilde{Y}_{f} \tag{3-17}$$

Functional representation of household behavior

After the income has been distributed, it is now relevant to ask how it is spent. As indicated above, the circular flow of income in the economy results in income collected by private households. This income can be used for either consumption or savings. In general, households can consume commodities from each sector specified in the model. The households' demand for various products (C_i) is determined by maximizing utility with a Cobb-Douglas function under the restriction of the budget constraint. This yields constant budget expenditure shares for all commodities over all income groups, which coincides with the implicit assumption of an income elasticity of 1. This simplification is acceptable as distributive matters are not the focal point of the analyses to be carried out in this study. The value of consumption of households depends on the price (P_i^q) and the quantity consumed from each sector (C_i). The total value of consumption by definition has to equal the sum of income earned from all sectors, which is determined by totaling the sector earnings and deducting the mandatory payments households have to make for taxes and the share of total nominal income the households save:

$$P_i^q \cdot C_i=\Sigma_p \beta_{ip} \cdot \widetilde{Y}_p \cdot (1-t_p)\cdot(1-s_p) \tag{3-18}$$

Hence, another central variable of the model, the price for the composite consumption good, has to be determined. The distinction between domestic and imported goods means that there are potentially two different goods in each sector, each with different prices (P_i^d and P_i^m) and being consumed in different quantities (D_i and M_i). The price of these two goods is weighted with the respective demand quantities, which yields the average sectoral price of the composite consumption good:

$$P_i^q = \frac{P_i^d \cdot D_i + P_i^m \cdot M_i}{Q_i} \tag{3-19}$$

Finally, the fact that households in the economy save part of their income has to be represented formally. The total amount of savings (S) is determined by the fixed marginal propensity to save (s) times the remaining income after household taxes have been paid:

$$\tilde{S}_p = \Sigma_p s_p \cdot (1 - t_p) \cdot \tilde{Y}_p \tag{3-20}$$

Functional representation of transactions with the rest of the world

Liberalization of the Russian economy was one of the features intrinsic to the transition process. The foreign trade system has been completely restructured and features of the former centrally planned system such as state trade agencies and exchange rate rationing have been dismantled. In fact, the opening up of the Russian economy to foreign trade and the full convertibility of the ruble were examples of relatively radical reform measures. Hence, today, foreign trade is an integral part of the Russian economy and the organizational designs of marketing channels for exports as well as for imports were created early in the reform period at a much faster rate than marketing channels for domestic products.

At the same time, the FSU lost its position as the major player it used to be in some world markets prior to transition – because of the disintegration and because of the strong overall output decline in its successor states. This development particularly holds for Russia. On many product markets, Russia can be considered to be a *small country* with respect to its relative share of world markets of most tradables. The opposite is true, for instance, with respect to world raw material markets on which Russia is still an important and major player.

Notwithstanding the fact that Russia's trade regime is far from being consistently applied and fully complying with international standards such as those set by the WTO, the opening up of its economy has exposed Russia to foreign competition, capital movements, and world market prices. Hence, the modeling of Russia's external economic relations has to be a central feature of the model. The functional representation of foreign relations is given with the following equations.

Export supply is often modeled endogenously as the difference between domestic supply and domestic demand. Therefore, increasing domestic prices of a commodity (i) normally results in increased export supply, as domestic consumers reduce demand while producers of the commodity have incentives to produce more. In such cases, the price elasticity of export supply is determined by the price elasticity of domestic supply and demand. As long as there is no differentiation between the domestic and the foreign product, this specification is acceptable. Problems arise, however, when models comprise firms which produce exclusively either for the domestic or for the foreign market. This is the case if sectors are still highly aggregated. In that case, a sector such as 'manufacturing' might entail the production of machines which are produced *either* for domestic *or* for foreign markets. In such cases, the above-mentioned quantification of export supply can overestimate the supply response caused by average price changes in one sector which do not differentiate between the prices in both market segments. In a similar way to the discussion of product differentiation on the import side (see above), the development of Russia's economy in the transition period implies that this distinction is made in the Russia model. The Russian energy sector is one example of an industry in which a clear distinction between the 'commodity' supplied to domestic markets and 'the one' delivered to international markets is being made (IVANOVA and PAVLOV 1999). Hence, export supply in the Russia model is c.p. dependent on the split of the production in sector i in the reference period (δ_i^t), the relative price between the products of sector i for exports (P_i^e) and the products price on the domestic market (P_i^d) as well as the transformation parameter (σ_i^t):

$$E_i^S = D_i \left[\frac{(1-\delta_i^t)}{\delta_i^t} \frac{P_i^e}{P_i^d} \right]^{\sigma_i^t} \quad where \, \sigma_i^t = \frac{1}{\sigma_i^t - 1} \tag{3-21}$$

As indicated above (eq. 3–4), producers split the total value of production in each sector ($P_i^x \cdot X_i$) in one component for domestic sales ($P_i^d \cdot D_i$) and one for export supply ($P_i^e \cdot E_i$). The sectoral price of exports depends on the world market price of the respective tradable (P_i^e), the export tax rate (t_i^e), and the exchange rate:

$$P_i^e = P_i^{\$e} (1 + t_i^e) R \tag{3-22}$$

Export demand (E_i^D) is then dependent on the exogenously given amount of exports in the specific sector. Additionally, the relative price in this equation reflects the following relation: while the price of the exported commodity on the world market is exogenous in the model, the domestic export price given in foreign currency and free on board can change because of shocks to which the model is exposed. In fact, the equation above indicates that the domestic price for sectoral exports (P_i^e) is linearly dependent on the fob-price of the same commodity in foreign currency ($P_i^{\$e}$), which again is determined endogenously as a result of export demand and supply. This specification does not allow that a price increase for the export commodity necessarily results in a complete loss of the sector's export market nor that domestic producers can dispose of arbitrary amounts of the exportable on world markets by simply lowering the product price. Instead, the more realistic assumption of a price elastic export demand is modeled by including the relevant elasticity (η_i^t) in the following equation:

$$E_i^D = \overline{E}_i (P_i^{\$e} / \overline{P}_i^{\$e})^{-\eta_i^t} \tag{3-23}$$

Now, the import side of the model has to be specified. Two equations are essential for this: one for import prices and one for import demand. The price for the imports depends on the exogenously determined import price ($\overline{P_i^m}$), the import tariff (t_i^m) and the exchange rate (R):

$$P_i^m = \overline{P}_i^m (1 + t_i^m) R \tag{3-24}$$

The fact that in the above function the import price is exogenously determined indicates that Russia faces fully elastic import supply. This relates to the above-mentioned fact that we modeled Russia on the import side as a 'small economy' which, at first sight, might be surprising with respect to the country's agro-food trade. Indeed, prior to transition, food and agricultural imports of the FSU consisted of only a few products. Cereal imports were not only substantial because of the need for feed in the highly subsidized domestic livestock sector but also constituted high shares of total world imports and, hence, had the potential to impinge on world market prices. After transition began and direct subsidies for food commodities had declined significantly, the structure of food imports diversified. Russia's imports of agricultural raw materials declined and were substituted by a vast number of imported processed food products, many of which originated from western industrialized countries. Nowadays, agro-food imports to Russia are much more diversified. The country's share in total world agro-food imports only comprised 2.8% in 1997 and dropped even further to 2.2% after the devaluation of the ruble following the financial crisis which hit Russia after mid-August 1998 (FAO 2000). Russia's relatively small share in total world imports also applies for all other major tradables. Hence, the *small country assumption* is modeled consistently on the import side.

Following this argumentation, import demand is determined as a function of total domestic demand for the sectoral good (D_i), the relative price between the domestically produced and the imported good (P_i^d/P_i^m), the split of total consumption in the base run situation (δ_i^q), and the elasticities of substitution:

$$M_i = D_i [\frac{\delta_i^q)}{(1-\delta_i^q) P_i^m} \frac{P_i^d}{}]^{\sigma_i^q} \; with \sigma_i^q = \frac{1}{1+\rho_i^q} \tag{3-25}$$

This implies that import demand in the model is sensitive to relative changes between prices for the domestic and the imported goods. Having specified import demand, total commodity demand (Q_i) is represented as the result of a CES aggregation function, which specifies the sectoral rate of substitution between the domestic and the imported commodity that can be induced by relative price changes. The function captures the fact that domestic agents demanding the sectoral good minimize the purchasing costs for the relevant commodity by substituting between the imported (M_i) and domestic good (D_i) accordingly. Again, the a_i and δ_i are scaling and distribution parameters, respectively, to represent the split between imports and goods of domestic origin in the initial situation, while the substitution parameter is represented by ρ_i.:

$$Q_i = a_i^q [\delta_i^q M_i^{-\rho_i^q} + (1-\delta_i^q) D_i^{-\rho_i^q}]^{-1/\rho_i^q} \tag{3-26}$$

Functional representation of government behavior

As indicated in Figure 3-4, the third important institution in the model is the government. No distinction is made in this model between the three different tiers of government in Russia. Instead, all officially recorded economic transactions of local, regional, and the federal Russian government are represented by one government agency whose behavior is modeled with a set of nine equations. They capture the different financial activities of the government by specifying six different sources of government revenues, one equation showing the total resources and three equations defining the expenditure side of the government budget. In general, the expenditures of the Russian government are used for consumption, export subsidies, and for government savings. On the revenue side, a distinction is made between various sources of potential government revenues accruing from trade activities and domestic taxation.

Despite the initial liberalization in the transition period, customs tariffs have been raised in the course of transition again. Particularly in the light of the increasing budget deficit, import tariffs have been an important source of revenues. In the respective equation, the total amount of tariff revenues (T^m) depends on the sectoral customs tariff rates (t_i^m), the world market price of the imported good (P_i^m), the exchange rate (R), and the quantity of sectoral imports (M_i). Again, total tariff revenues can be obtained by totaling over all sectors:

$$\widetilde{T}^m = \Sigma_i t_i^m \cdot P_i^m \cdot R \cdot M_i \tag{3-27}$$

Similarly, revenues from export taxes can be specified. In this case, it is the quantity (E_i) and the price of the export good (P_i^e) which has to be multiplied by the exchange rate and the sector-specific tariff rate (t_i^e). However, note that the export tax rate can also be negative, which would turn the term negative, indicating that the government has 'negative' revenues, i.e. expenditures. In fact, this will be one of the empirical features of the Russia model, since export subsidies have been one way of transferring subsidies to specific sectors (see chapter 5):

$$\widetilde{T}^e = \Sigma_i t_i^e \cdot P_i^e \cdot R \cdot E_i \tag{3-28}$$

On top of revenues from trade activities, the government has various alternatives to tax households and firms. Indeed, the Russian tax system has been changed rapidly to resemble a western type tax structure. The decentralization policy of the first transition decade has also resulted in the introduction of various taxes on the regional and local level. Various tax exemptions for Russia's agro-food sector were introduced to compensate farmers and food processing firms for the loss of direct subsidies experienced during the early transition years. The most important types of taxes have been specified in another version of the Russia model which has been used for tax policy analysis (WEHRHEIM and WIEBELT 1997a, 1997b and 1998). However, in this study not the whole range of taxes used in the Russian Federation needs to be specified as the focus of this study is not on tax policy reforms. Instead, to trace the differing economic effects of various taxes, only sales taxes and, hence, indirect taxes were specified on top of the 'taxes' levied on imports and exports.

To model sales taxes (T^x), the sector-specific consumption tax rate (t_j^x) has to be multiplied by the value of consumption in the respective sector:

$$\widetilde{T}^x = \Sigma_j t_j^x \cdot P_j^x \cdot X_j \tag{3-29}$$

What follows is the aggregation of government revenues into one function by adding all six sources of government revenue:

$$\widetilde{Y}^G = \widetilde{T}^m + \widetilde{T}^e + \widetilde{T}^x \tag{3-30}$$

Note that all forms of government revenue could also assume negative values, which would reflect government expenditures or subsidies. This could become relevant, for instance, when one intends to exogenously introduce export subsidies in the agricultural sector which actually will be one of the simulations in Chapter 5. However, having this option, we only have to define direct government consumption (C_i^G) to complete the expenditure side of the model. This is dependent on pre-determined and fixed shares of government consumption in each sector (β_i^G), multiplied by the exogenously determined total level of real government consumption.

$$C_i^G = \beta_i^G \cdot \overline{C}^G \tag{3-31}$$

The remaining function represents another important identity, namely the government budget balance (S^G), which is the difference between total revenues (Y^G) and the value of government consumption ($P_i^q \cdot C_i^G$) totaled over all sectors. The budget is balanced via the capital account, which basically represents a 'National Bank' receiving either the government budget surplus (savings) or financing its deficit (debts):

$$\widetilde{S}^G = \widetilde{Y}^G - \Sigma_i P_i^q \cdot C_i^G \tag{3-32}$$

Gross Domestic Product

Applied CGEs are often called micro-macro models because they model both aspects of the economy. So far we have discussed the micro behavior of economic agents in the model. In order to ensure that the model is also consistent with macroeconomic theory, we now have to define several macro identities and apply them to the Russian economy. First, nominal and real GDP have to be specified. Nominal GDP is calculated at factor costs (GDP^n) by totaling the value added over all sectors ($P_i^v \cdot X_i$), adding indirect, import tariffs (T^m), and substracting (adding) export taxes (subsidies) (T^e):

$$GDP^n = \Sigma_i P_i^v \cdot X_i + \widetilde{T}^x + \widetilde{T}^m - \widetilde{T}^e \tag{3-33}$$

Real GDP (GDP^r) is calculated at market prices by totaling the various components of final demand in each sector. Hence, private (C_i) and government consumption (G_i) enter into the calculation of GDP; secondly, demand for investment goods (I_i^A) and demand for change in stocks (I_i^W). Finally, the balance of trade (the value of exports minus imports denominated in world market prices multiplied by the exchange rate) has to be included in the calculation of GDP:

$$GDP^r = \Sigma_i (C_i + C_i^G + I_i^A + I_i^W + E_i - \overline{P}_i^{\$m} \cdot M_i \cdot R) \qquad (3\text{-}34)$$

The definition of GDP serves another important additional function. As a money market is not specified in the Russia model, only relative prices and no absolute prices are defined within the model. Hence, all prices have to be specified relative to an exogenously defined price. Here, we define a GDP deflator as the *numeraire*. The exogenously determined price index is therefore the relation between nominal and real GDP:

$$\overline{P} = GDP^n / GDP^r \qquad (3\text{-}35)$$

Savings

The capital account collects not only government savings (S^G) but also total domestic savings (S) which comprises private savings from households (S_p), depreciation and foreign savings (S^F). Again, if the capital account is conceptualized as a national bank, it is easy to understand that any savings could also be negative, thus representing debts to the capital account. If foreign savings are negative, this would represent a case in which a share of total savings is transferred abroad, because of either higher relative rates of return received for savings abroad or because high disincentives to keep savings on domestic accounts induce capital flight:

$$\widetilde{S} = \widetilde{S}_p + \widetilde{S}^G + \widetilde{A} + \widetilde{S}^F \qquad (3\text{-}36)$$

Closure of the model

In a final step, the equilibrium conditions of the model have to be defined. The model is 'closed' by actually defining different balances which have to hold in the base run situation. Only if these balances are in equilibrium is the model calibrated correctly. There are a total of four closure rules which are used to solve the Russia model. The first one represents the fact that the model must find a solution in which all sectoral commodity markets are cleared, which basically represents a demand-supply equilibrium, irrespective of the source or use of supply and demand. Total supply in one sector (Q_i) has to equal total demand for the composite commodity in this sector. Hence, on the demand side, all sources of demand have to be added: demand for intermediate demand (V_i), household demand (C_i), government consumption (C_i^G), as well as investment demand (I_i^A) and demand for the formation of stocks (I_i^W):

$$Q_i = V_i + C_i + C_i^G + I_i^A + I_i^W \qquad (3\text{-}37)$$

The next closure rule assures that factor markets are in equilibrium. The supply of primary factors (F_j) is exogenously set and has to equal the demand for production factors totaled over all sectors:

$$\overline{F}_f = \Sigma_i F_{if} \tag{3-38}$$

Furthermore, the trade balance must be maintained. This can be captured by equalizing the value of imports ($\overline{P}^{\$m} \cdot Mi$) totaled over all sectors with the value of exports ($P^{\$e} \cdot Ei$) plus the foreign savings ($\overline{S}^F$), which again can be positive or negative:

$$\Sigma_i \overline{P}^{\$m} \cdot M_i = \Sigma_i P^{\$e} \cdot E_i + \overline{S}^F \tag{3-39}$$

The final closure rule of the model is the savings-investments identity which ensures that all investments made in the economy (I) are financed with savings (S), whatever kind they may be:

$$\widetilde{S} = \widetilde{I} \tag{3-40}$$

The equilibrium

The demand-supply balance in each sector of the model is found by a variation of prices. As outlined above, the model defines nine prices over $1 \ldots n$ sectors: the sectoral producer price (P_i^x), the net price (P_j^v), the capital goods price (P_j^k), the price of the composite good (P_i^q), and the price for the domestic good (P_i^d). To model foreign trade, the following prices are defined: export (P_i^e) and import prices (P_i^m) are given in domestic currency as well as in foreign currency. While the *fob* export price ($P_i^{e\$}$) is determined as the equilibrium between export supply and export demand, the sectoral import price is exogenously determined because Russia is modeled to be a 'small country' on the import side. Hence, deducting the two exogenously defined prices from the total number of nine prices, seven prices remain which have to be determined within the model. Out of these, however, six prices are linearly dependent on other prices (P_i^x, P_j^v, P_j^k, P_i^q, P_i^e, P_i^m) and, hence, on top of the *fob* export price ($P_i^{\$e}$) only the price for the home good (P_i^d) has to be determined in each sector endogenously.

Tables A3-1 and A3-2 (in the Appendix) summarize all equations of the model and provide an overview of all variables in the model and the respective notation. If the number of equations ($19 \cdot n + f \cdot n + 2 \cdot f + p + 15$) and endogenous variables ($19 \cdot n + f \cdot n + 2 \cdot f + p + 14$) are compared, it can be seen that the model contains one more equation than variables. However, one equation is linearly dependent on the other equations, which means that one equation can be dropped when determining the general equilibrium. This ensures that the number of equations in the model equals the number of variables and, hence, that Walras' second law holds. The choice of which equation is best to be dropped is open to discussion (cf. ROBINSON 1991). Often the identity between savings and investments is dropped and investments are determined endogenously as the sum of savings from all sources in the Russian economy.

The main features of the structural model version: representation of market imperfections in the model via alternative closure rules and other features

While CGE models are normally perceived as purely neoclassical devices of policy analysis, they can modified in such a way that they represent a structural economy as described in chapter 2. In fact, the structural features reflect some of the important market imperfections which can be introduced into the model with various mechanisms. In the following, we will discuss the motivation for taking these market imperfections into account and how they have been represented in the model.

In principle, in a well-developed market economy, various market institutions ensure that the flexibility of resources in the economy is high and that markets can adapt efficiently to relative price changes. In economies of transition, this is not always the case. Instead, market imperfections and structural rigidities hamper full adjustment to changes in relative prices. Poor institutions, the lack of information, and high market risks were factors contributing to poorly functioning and inefficient markets in Russia in the mid–1990s, and they are to date. To take these features of the contemporary Russian economy in the model into account, several structural features are incorporated into the model, which all aim to reduce the flexibility in the model economy. Obviously, care has to be taken in the design of such rigidities. If the model is designed in such a way that very few endogenous variables can adjust to policy shocks, the results might easily be misleading. In general, the flexibility within general equilibrium models can be reduced by fixing specific variables which are central for the closure of the model (e.g. investments) at their pre-determined level, by reducing the mobility of resources (e.g. capital or labor), by specifying elasticities at low levels and, in doing so, keeping the responsiveness of variables to changes in other variables weak (e.g. the elasticity of substitution between domestically produced and imported goods). All of these options are used in the Russia model and will be described in the following in more detail.

Degree of tradability. The experience of Russia's transition process has shown that one of the most unique features has been the restructuring of trade. Imports, for instance, of primary agricultural commodities such as cereals decreased, while imports of processed food products glutted domestic markets. Many theoretical models used to distinguish between 'traded' and 'non-traded' goods. Applied models instead focused on the notion of tradability as a 'continuum', rather than as the above-mentioned dichotomy. The degree of tradability in such applied models for transition economies should, therefore, be characterized by different elasticities, which is achieved using constant elasticities of substitution on the import side and constant elasticity of transformation on the export side (see chapter 4 for the absolute elasticity values chosen). Even though the distinction between 'traded' and 'non-traded' goods becomes increasingly obsolete due to this modeling feature in our model of the Russian economy, we also exclusively treat some sectors as 'non-tradable' sectors. For instance, small-scale agriculture, with its low degree of commercialization and no significant links with foreign markets whatsoever, is one obvious example of such a sector. This feature has one important implication: any foreign exchange-related exogenous shocks will have no direct effects on these non-tradable sectors. Only second-round effects, for instance changes in relative domestic prices that were induced by the initial effect, might then affect the tradable sectors. If a *non-tradable sector* is also widely insulated from other production sectors within an economy, such second-round effects might also be limited.

Factor mobility. Factor mobility between sectors in transition economies is limited. Weak factor mobility not only restricts internal price adjustments but also inhibits the ability to respond without friction to changing international prices. Hence, such limitations in factor mobility should be taken into consideration when modeling transitional economies. In fact, the functional form chosen to represent the basic behavior of economic agents can already serve to reduce the flexibility of the model. Despite the theoretical caveats of the fixed input representation on the production side, this might indeed be a relatively realistic representation, taking the long duration of restructuring in the Russian economy into account. An empirical analysis of labor market adjustment of medium- and large-scale firms in selected sectors of the Russian economy, for instance, reveals both low degrees of labor demand and low elasticities of substitution between labor and capital. In fact, one of the implications of the analysis is that substantial segments of the Russian economy are to be typified by a Leontief-type production behavior (KONINGS and LEHMANN 2000).

Labor market rigidities. Another structural feature characteristic of Russia's transition period is that relative real wage rates have been kept relatively stable. Sectors with traditionally high shares of employment are restructuring only slowly, and relative wage rates in sectors which reveal relatively low competitiveness (e.g. mining) are prevented from declining more strongly for social reasons. The strikes in the mining sector in spring 1998 which preceded Russia's financial crisis in mid–1998 are just one example of the formation of political opposition against further wage cuts in non-competitive sectors. Hence, in the Russia model, nominal wage rates in each sector have been fixed, at least in the standard closure. In selected experiments this restriction will be released to further identify its impact on the overall responsiveness of the model economy.

Capital mobility. One other striking feature of the transition process is the deterioration of the capital stock. In fact, it had been expected that a significant share of the capital stock originating from the era of central planning, which was generally more in favor of heavy industry and large-scale industrial plants than of service or high technology industries, could be converted to a more competitive industrial structure by privatizing it. Instead, however, the old capital stock became obsolete in most transition countries. Hence, the restructuring of the Russian economy did not coincide with significant inter-sectoral transfers of the capital stock, which means that the capital stock in each sector of the Russia model is kept fixed. As a result, changes in the prices for outputs, intermediates, and production factors will yield different profit rates in each sector. Indeed, the transition process has yielded substantial restructuring of the economy. To capture the reallocation of resources, output, and trade across sectors, a division between industry and agriculture as in stylized models may not be sufficient in order to distinguish between, for instance, consumer, intermediate, and capital goods. Here, the policy focus of the model should be decisive for the disaggregation chosen. Due to the explicit focus on the agro-food industries, 10 out of 20 sectors in the IOT are agro-food sectors, while other sectors have been disaggregated to identify the most important input sectors for the agro-food sectors.

Investments are fixed. As mentioned above, the maintenance of the savings-investment balance is one of the central macroeconomic 'closures' of the model. In order to solve the model and maintain this balance, either savings or investments must be fixed. Particularly in neoclassical models investments are mostly kept flexible. In our structural model world, however, we have kept the total amount of investments available in the economy fixed for two reasons: a modeling reason and an economic reason. First, the model we have specified is not a dynamic model in which the adaptation path of investments would be a central feature. Instead, the Russia model is a static-comparative one which is solved over the 'short run', which means we expect that total investments remain constant. Secondly, against the background of the Russian economy in the mid–1990s, the willingness to invest in the economy was weak, most likely due to the relatively high risk perceived by potential investors. In fact, the risk index shown in chapter 2 documents this, at least for foreign investors. Instead of being invested, additional income generated by households was saved, particularly by private households. On the contrary, despite the mal-performance of the Russian economy, invested capital could hardly be withdrawn from the 'real sectors'. Due to the specific sectoral uses of most investments, they could not be disinvested from any sector and monetized once the return on investment in the respective sector deteriorated.

In reality, investment-savings issues were obviously far more complicated. For example, household savings were mostly made in dollars and not in rubles. However, within the framework of our model, we can only represent this relationship in a stylized fashion. Therefore, we defined that the *marginal propensity to save of households adapts instead of investments* in order to keep the savings-investment-identity balanced.

Neoclassical versus structural economy. Due to all of these structural rigidities described, not all resources available in the economy can adapt to exogenous changes to which the Russian model economy was exposed in a way it would be expected from a neoclassical model. In fact, some resources can actually remain *unemployed.* For instance, the standard closure of the labor market will allow for excessive supply of labor and fixed wage rates, allowing unemployment to occur. In contrast, additional labor can be employed if, for instance, an exogenously determined positive shift in total factor productivity enhances the productive capacity. This reduces the pressure to restructure within the economy. Clearly, these features distinguish this model from a *neoclassical* model in which the solution would always have to be found along the production possibility frontier.

As a consequence of these modifications, our model is typified as a *structural model*, as described in chapter 2. The model economy can operate below or above the production possibility frontier and policy experiments can yield both outward or inward shifts within the economy (see Figures 2-6 and 2-7). Adaptation to exogenous shocks is limited to specific factors in the economy, which might result in stronger responses of individual indicators. At the same time, the responses might be different from what might be expected at first sight with neoclassical theory in mind. Before we discuss the effects of this model on simulation results, we first need to elaborate the second major component of this modeling exercise, the data base, which is the topic of the next chapter.

Chapter 4

The Empirical Version
of the Russia Model

The following chapter discusses the data base of the CGE model for Russia. In doing so, we will discuss the various steps involved in compiling a consistent set of data for a CGE model, touch upon the associated difficulties for an economy that is in transition, and elaborate the procedure required to update the data base from one reference period to another. The objective is to combine a consistent data base with the theoretical model described in the previous chapter, thereby developing an applied model for the entire Russian economy. With this model version, we will then be in a position to look into the effects of exogenous shocks (e.g. changes in economic conditions or policies). By carrying out economic and policy simulations, we will be in a position to assess the allocative effects of such experiments quantitatively.

The general procedure in compiling the data base for an applied CGE model

The previous sections indicated the wealth of statistical information required to complement the theoretical framework of the CGE model with a *consistent numerical representation* resembling reality as closely as possible. For each endogenous variable in the model, a value for the base run of the model has to be specified. The large number of variables identified in the model (Table A3-2) makes the need for efficient data management tools obvious. Therefore, in order to handle the data set for the Russia model, two closely related *data management tools*, both with a firm theoretical rooting, will be used:[1]

* an Input-Output Table (IOT) and
* a Social Accounting Matrix (SAM).

As for all applied CGE models, the *SAM* is the centerpiece of the empirical analysis, as it provides a consistent framework and the structure for tracing

---

[1] There are various unique discussions of social accounting matrices and input output tables as data tools for economic analysis. One classical study is PYATT and ROUND (1985), in which the methodology of social accounting is described, various examples of SAMs for developing countries are given, and SAM-based models are elaborated on. BULMER-THOMAS (1982) and HELMSTÄDTER et al. (1983) describe the various means of input-output analysis as a tool for research on the structure of an economy. SADOULET and DEJANVRY (1995) provide a discussion of the sources, methods, and applications of input-output analysis in developing countries.

economic transactions in an economy at a given point in time. It reveals the *circular flow of all economic activities* within a reference period, normally a calendar year, in the respective region, i.e. the flow of money, factors and commodities in the Russian economy. The SAM contains an IOT that shows all inter-sectoral links, the links between industrial sectors, and non-industrial factors of demand and value added. Hence, *the IOT is a subset of the SAM*. The concept of the IOT dates back to the work of the Russian economist Vassilij Leontief, whose main achievement was '... to provide an empirically implementable general equilibrium system' (BULMER-THOMAS 1982: 55).

The *major economic actors* usually shown in a SAM are the same as those that were listed in Figure 3-3 as the major institutional players in the Russia model: enterprises, the government, households, and third countries, or the rest of the world. The different economic activities of these actors are obviously linked with one other, and this is also revealed by the SAM: producers use goods from other sectors of the economy as inputs in the production process, employ labor from households for which they pay wages and earn profits. Households earn income, spend it on consumption, pay taxes to the government or save. Government uses its revenues to finance public investments by buying goods from the commodities account or subsidizing selected sectors in the economy at the expense of others. It can also have savings or debts. The rest of the world can sell imports as intermediates or for final consumption, receive exports and can be linked to the domestic economy further via a net capital inflow or outflow. Hence, the SAM empirically reveals the circular flow of income in an economy as it has been formulated theoretically in the previous chapter.

When speaking about a SAM, it is useful to make a distinction between what could be called a *macro-* as opposed to a *micro-SAM*. The former concept contains a complete picture of the circular flow of income in an economy based on macro-totals only. As indicated in Table A4-1 (in the Appendix), these macro-totals are 'booked' in a specific account, each representing one of the institutions in the economy. In this example, two accounts represent the firm sector (activities and commodities) and one account each represents the remaining three major institutional actors that are distinct from one another in this model.

The overall *concept of the SAM* is based on *double-entry book-keeping*, as for each of the institutional actors two accounts are defined. The value of the respective economic indicator is 'booked' twice: on the one hand as expenditures (across columns) and on the other hand as revenues (across rows).

The firm sector is represented by two accounts: the *activities* and *commodities* account, both of which are used to better *trace the inter-industry flows*. While the former can be perceived as the *production account*, the latter can best be understood by picturing it as a *wholesale market* that buys goods from the *activities* (domestic sales) and the *rest of the world* (imports) and sells these products again to *activities* (intermediate demand) and the various other final demand components. Along the *commodities'* row account goods are demanded as intermediate inputs, for private and government consumption as well as for investments. The first two accounts describe transactions of *n* economic sectors for domestic as well as for imported and exported goods. Hence, the very upper left-hand corner of the SAM also represents the input-output part of the economy.

All other accounts in the SAM reveal the *circular flow of income*: the origin of income on the production side is shown as well as the uses of income for consumption and savings. The *factor accounts,* for labor and capital, receive payments for the *creation of value added* (across rows) and then have to redistribute their earnings as profits, wages and factor taxes to households and to the government account (across columns). In a next step, this factor income is redistributed as institutional income: *households* receive profits and wages and the *government* levies taxes on both of these; retained profits are transferred to the *capital account*. Hence, the row account for the *government* shows several revenue sources: indirect taxes from sales of commodities, revenues from import tariffs, and direct taxes levied on wage and capital income. The column account again shows the 'costs of running' the *government*: expenditures can be made to subsidize exports (negative payments would indicate export taxes), to finance the provision of public services and to finance the government budget surplus (positive payment) or budget deficit (a negative payment). The receipts of the *capital account* across rows therefore show total savings of all economic actors: saved profits, household savings, and the budget deficit or surplus. Finally, the account for the *rest of the world* indicates the export and import activities as well as the trade balance, which again is booked in the *capital account balance*.

Based on the macro-SAM, a *micro-SAM* can be constructed. It differs from the macro-SAM in three major ways: first, the very upper left-hand corner of the macro-SAM is disaggregated by revealing all details about the inter-industry flows between all sectors specified in the Russia model and represents the IOT. Second, depending on the interest of the study, the institutional accounts can be further split up in the micro-SAM, which is one of the attractive features of this approach. The government account, for instance, could be split into the various levels of government. In the case of Russia, federal, regional, and local government accounts could be distinct from one another and additional accounts for other extra budgetary agencies such as social security agencies could be specified. Alternatively, households could be split up into rural and urban households and/or by income groups. However, to do so precise data is required for *all* economic transactions of the respective sub-groups of each institutional account with all other economic agents in the model with which economic relations exist.[2] Third, all of the macroeconomic totals given in the upper right-hand corner and in the lower left-hand corner of the stylized SAM have to be disaggregated according to the sectoral split chosen for the respective model. For instance, the total sum of government expenditures has to be split up among the various sectors of the economy according to the actual government consumption. The distribution of government expenditures among the various sectors would then be written in one column under the heading of the government. In a similar way, the macro-totals given in the lower left-hand corner of the matrix would be disaggregated sectorally. However, in this case, sectoral disaggregation is revealed row by row.

---

[2]   In fact, this necessity explains why the four major economic agents in the Russia model will not be split up further.

Having done so, the matrix still needs to be balanced from the expenditure and revenue side. On the one hand, expenditures for intermediate demand totaled over all sectors, plus all payments for value added components (labor, capital, tariffs), plus imports and import tariffs have to equal total domestic expenditures. On the other hand, the revenues from the final demand components have to be added to the revenues from sales of intermediates (again totaled over all sectors) to obtain total domestic revenues. This balance must hold for each individual sector and for the total of all sectors.

To manage these data sources properly, it is helpful to design a spreadsheet resembling the structure chosen for the micro-SAM. Here, it should be clear that many options for disaggregating the micro-SAM exists. Basically, each cell in the macro-SAM could be split up and could constitute another sub-matrix in the micro-SAM. When compiling the micro-SAM, two constraints have to be considered: first, empirical values for each variable defined in the theoretical model are mandatory and, second, the micro-SAM must be balanced, that is, various equilibria must be maintained.[3]

One example of a micro-SAM is shown in Table A4-2. It is based on the fact that the SAM must be square. The underlying equilibrium is the *equality between the value of domestic production calculated from the cost side* (earnings approach) and *the value of domestic production calculated from its use side* (product approach).

The two approaches are summarized in Table 4-1. While the first concept totals all expenditures and, hence, cost components of each sector in the economy, the second one totals the sectoral revenues for selling products to the various sources of economic use. Using the *cost approach*, the total costs for intermediates used in the production process have to be identified first. Value added components then have to be added, including payments for capital (e.g. interest rate payments, depreciation, stocks, or distributed earnings) as well as those for wages. Indirect tax payments also have to be added to obtain the total value of domestic production in cost terms.

The calculation of domestic production using the *product approach* necessitates the identification of the components of total demand for domestic production. Again, the starting point is the total sum of intermediate demand for which the economic sectors receive revenues. Then, various final demand components have to be added, such as household consumption, government consumption, and investment demand. As we are interested in the total value of domestic production, we have to subtract from total domestic absorption the value of imports (including customs tariffs) and add the value of exports (add or subtract export taxes or export subsidies, respectively).

This equilibrium is closely related to the calculation of GDP which can be estimated both from the production or from the demand side. *GDP at factor costs* incorporates capital costs, labor remuneration, indirect taxes and revenues from foreign trade (import tariffs and export taxes). It equals *GDP at market prices*, which

---

[3] SHOVEN and WHALLEY (1992: 110) give a whole list of potential 'consistency checks' which can be used when compiling a benchmark equilibrium data set: demand equals supply for all products; total costs equal total sales in each industry; the endowments of consumers match factor usage; the value of final demand equals the sum of value added (etc.).

is the sum of all final demand components, including private consumption, government expenditures, the balance of trade, and investment demand. This equality generally must hold not only for the economy as a whole but also for each individual sector, and can be empirically shown in the micro-SAM. In fact, the micro-SAM in Table A4-2 has been constructed on the basis of this equality. However, the components are split up by sectors, yielding further insight into the economy. For instance, while the sector-specific sums of expenditures and revenues for intermediates do not have to be equal, one first check of the input-output part of the micro-SAM is that the sum of all expenditures and that of total outlays must be equal.

Table 4-1   Calculation of total domestic production using the cost approach and the product approach

Origin of domestic production (cost approach)	Use of domestic production (product approach)
Expenditures for	*Revenues from sales of*
intermediates + capital (interest payments, depreciation, stock building, distributed earnings etc.) + wages + indirect taxes	intermediates + goods to private households + goods for government consumption - imports (incl. tariffs) + exports (incl. export taxes)
= Total costs of domestic production	**= Total use of domestic production**

After these qualifications about the two major elements of the model's data base, Figure 4-1 provides an overview of the steps required to compile a fully consistent data base for the economy-wide model. While it makes sense to carry out most of these steps consecutively, some might be implemented in parallel, and feedback loops might be unavoidable to enhance the overall consistency of the data base.

The compilation of the empirical version of the model normally starts with the *search for the most recent input-output table* and *macro data* for a given year. In an ideal case, data is obtained for both components from the same source. That would ensure the highest data consistency possible but is, at the same time, rare. The tedious work involved in compiling a fully balanced and fairly disaggregated IOT means that they are normally only compiled on a rather irregular basis. Hence, data from different sources often has to be combined, which can create particular difficulties in the case of transition economies for which the methodology in collecting the relevant data might not yet be consistent with internationally accepted standards. Therefore, one task is to ensure *compatibility* of input-output data and the macro-data by checking that, for instance, the total value of intermediates given in the input-output table corresponds to that from national accounting data. Another major step may be the need to *update the IOT* if the data for the most recent IOT is relatively outdated. This is particularly relevant for transition economies which have experienced significant restructuring, implying also changes in the structure of forward and backward linkages in the economy. Alternative methods of updating exist, and these can be applied if an older IOT can be used as a starting point. No matter which method is used, the dominating rule should be to integrate as much information for the most recent base period as is available. Hence, it is necessary to return to the compilation of consistent input-output data and then go to all the effort of balancing the material again.

Once a fully balanced IOT and a macro-SAM have been compiled, the next task is the *disaggregation of macro totals* (e.g. total household consumption) by sector. For this step, additional data is required, and this can originate from macro- and/or micro-economic data sources.

In fact, most national accounts do not provide sufficient information about the sectoral decomposition of macro totals, which means that the *combination of macro and micro data* is often mandatory. Further disaggregation of accounts for specific economic agents actually necessitates the use of primary data. Disaggregation of households by income groups or other socio-economic characteristics, for instance, could take place with data from available nationwide surveys or data collected via small random samples.

On top of the disaggregation of macro-totals, another step might be to *disaggregate the economic sectors* in the economy further. Particularly if the focus of interest is a specific sector of the economy, the data available from the official IOT might not be sufficient. In fact, a focus on the agro-food sector often necessitates further disaggregation of this sector into various sub-sectors, for instance plant and livestock production, in order to obtain tangible results of different policy simulations relating to policy instruments that are relevant for specific production sectors only. Equally important is the disaggregation between primary production and food processing because of the different technologies applied in these sectors.

When the data has been collected, it again is essential to insert it into the micro-SAM in such a way that all accounts are balanced. Once a balanced micro-SAM has been set up, the *input-output coefficients* (IOC) can be calculated. Furthermore, all policy variables, shares, and elasticities have to be exogenously determined and inserted into the data base. Together, these are the elements which have to be linked with the theoretical model version. Due to the constraints formulated in the theoretical model, the remaining parameters (e.g. efficiency parameters) can then be calculated correctly, i.e. without any leakage, if the data set is complete and consistent.

Data issues linked with the compilation of a Social Accounting Matrix for an economy in transition

CGE analysis of transforming economies is still in an early stage (see section 3.2). Frequently, data problems are a first obstacle to following this modeling approach. The previous section gave an indication of the variety of issues that have to be solved when compiling the data base for any country. We will now turn to the country-specific issues and, hence, the situation in Russia.

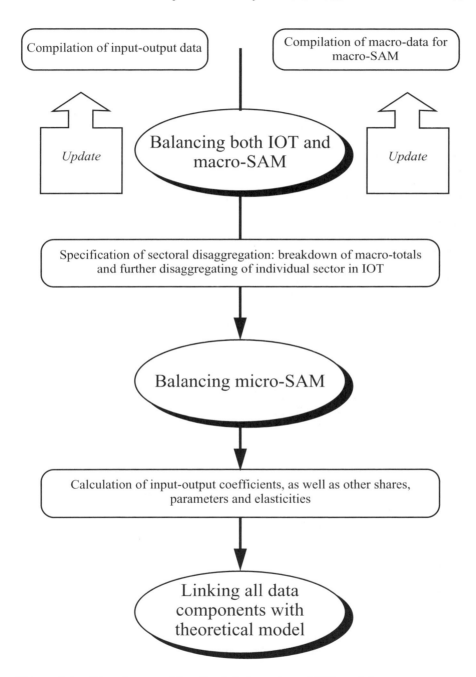

Figure 4-1   Steps in compiling the data base for a CGE model

The fact that up to this point *no fully consistent and highly disaggregated IOT for Russia* for any given year in the transition period which would be based on the System of National Accounts (SNA) has yet been published by either the Russian statistical office (GOSKOMSTAT) or international institutions such as the World Bank or the IMF is an indication of the surmounting problems at hand. Indeed, there are several reasons why this task is particularly daunting in the case of Russia (and in the case of other FSU countries): first, after the break-up of the FSU and the revision of the statistical system, it became more difficult to compile national accounting systems for the newly independent states because of the structural break associated with the disintegration process. Second, not all restructuring which went along with the transition from plan to market took place at an observable level. Significant levels of capital flight, high shares of informal and even criminal activities are not represented at all in Russia's national accounts. In fact, one explanation for the particularly strong output decline in the Russian economy during the first years of transition was that a high share of economic activities shifted towards the *hidden economy*. Alternative estimates of output decline based on the change in energy consumption indicated that the output decline was indeed much less severe than expected. Third, even though the transition process which originated from one of Gorbatchov's major political ideas, namely *Glasnost*, has had substantial impact on the concept of national statistics, it is still difficult to obtain data on national production based on the SNA. Statistical information on agricultural production is a good case in point. During the Soviet era, figures on grain production were top secret. While this policy has changed, there is still great caution on the part of statistical bodies in the Russian Federation in publishing or releasing related data such as information on cereal stocks, inter-regional grain flows etc. At the same time, this caution is caused by the concern about the 'political correctness' of released data, mixed with great uncertainty about the reliability of collected data because of reasons related to the second point above.

The problems in finding adequate data for the FSU and its former member states were even more severe during the Cold War, when the primary objective of national accounting data was to serve political purposes instead of economic analysis. In the light of these difficulties in obtaining reliable data for the FSU, the World Bank commissioned two independent compilations of national account estimates for the republics of the FSU. One of these attempts resulted in a first compilation of IOTs for all CIS states which comply with SNA format (STEINBERG 1992). It was complemented by a WORLD BANK (1992) publication and later by a study commissioned by the OECD (1993), both addressing in a more detailed fashion the statistical and methodological issues in adapting the national accounts for the members of the FSU to the SNA format. These efforts yielded relatively detailed IOTs, containing between 100 and 125 sectors, for the CIS countries for the year 1990 (WORLD BANK 1995).

The adoption of the data to western standards was mandatory because the IOTs for the former centrally planned economies had several distinct features making it difficult to use them for CGE and other forms of economic analysis used in the western world. Multisectoral models which were frequently based on such IOTs have a long tradition in former centrally planned economies. As quantitative production schedules were most important, input-output models based on Leontief-type structures were also frequently used to address problems of quantitative planning in the economy.[4] In doing so, IOTs based on quantities and not on values were applied. Despite the problems associated with obtaining secret national account data for the FSU before transition started, Steinberg managed to obtain original IOTs for Russia and other republics of the FSU from the statistical office of the FSU. Using his knowledge about the Russian economy and supplementary Russian statistics, e.g. on financial issues, he managed to compile 18 sector IOTs for all republics of the FSU – the first for these countries based on the SNA approach (STEINBERG 1992). Some of the major discrepancies between these IOTs and the western SNA format are the following:

Net Material Product. One of the central issues in converting IOTs for former centrally planned economies into the SNA format was linked to the fact that national account data in these countries was based on a different statistical approach called the Net Material Product (NMP). This concept included the value of all goods being produced in the economy and services which were directly linked with the exchange of these goods. The computation of the NMP was from the production side only (TROSCHKE and VINCENTZ 1995: 14). Other services which were not directly linked with productive resources such as services offered by banks, insurances, universities or defense were not registered as components of the NMP. In order to bridge national accounts data based on the NMP to GDP, major adjustments were made for the 1987 IOT for Russia, such as inclusion of depreciation and the bulk of all officially registered services (WORLD BANK 1992).

Foreign trade. Official IOTs for the republics of the FSU excluded foreign trade revenues, as these were based on the NMP concept. Therefore, one other important task in compiling IOTs based on the SNA format was to correctly reveal trade flows. However, even though GOSKOMSTAT published trade figures for 1990, several computational changes had to be made. First, former internal trade between Soviet republics had to be counted as external trade. After the apportioning of extra FSU trade among the 15 republics of the CIS, trade among the subnational units had to be reclassified as cross-border transactions. Secondly, and even more important, the compilation of trade data was difficult because of dual ruble pricing, an accounting artifice for dealing with the inconvertibility of the old internal ruble into foreign currency. Cross-border transactions were first recorded in foreign currency and then at internal ruble prices and at foreign trade prices. Firms which delivered any products to the official foreign trade organizations reported the value of their sales to theses agencies in domestic ruble prices. Hence, trade was evaluated in domestic and in world market prices, both in rubles. Domestic trade firms were paid at domestic prices, but statistically all international transactions were evaluated at world market prices. The difference was attributed in the IOT to an artificial trade-agency-sector. The IOT compiled by Steinberg took this accounting artifice into

---

[4] In fact, the possibility to use the input-output approach for qualifying various government interventions was responsible for the resistance in using this approach in the US in the 1940s and 1950s as advocated by Leontief.

consideration and evaluated trade at foreign trade prices. Using this method, in 1987–90, Russia was the only net exporter among the FSU republics (STEINBERG 1992: 16).

Gray market activities. Furthermore, gray market activities and subsistence production were not or only insufficiently included in the calculation of the NMP. Criminal activities can per definition not be included in official statistics, despite their importance in the Russian economy. Some sector activities which are missing from national accounts data and IOTs were available from other secondary data sources, such as household income and expenditures data. For the early 1990s, the share of household expenditures in Russia and the CIS being channeled through gray markets was estimated at 40% (STEINBERG 1992: 2). The importance of these informal markets has been growing further in Russia in the transition period.

Service sector. The growing service sector was not sufficiently represented in statistical surveys of the FSU, either. Steinberg included information on the service sector from official data, which reveals a sharp increase of this sector as early as the late 1980s.

Self-subsistence production. In the course of transition, the importance of small-scale production of agricultural commodities and their informal sales increased in Russia as many of the large-scale collective farms, processing and wholesale institutions collapsed or experienced severe output decline. About 80% of Russian households engage in some form of subsistence production (THO SEETH 1997). In 1995, about 55 million households in Russia had access to private plots, where most of them produce vegetables, potatoes etc. but also keep livestock. The evaluation and integration of this informal sector in a current Russian IOT is therefore desirable, but linked with various difficulties. One approach to tackling the respective changes in informal markets is to extract the mandatory data for the IOT from additional primary sources.

Decentralization of policy-making. Even though some important methodological changes have been made by compiling new national account data for Russia, there remain data issues related to genuine economic and political changes in the transition period. For instance, as regional governments have gained independence, particularly with respect to agricultural and food policy-making (MELYUKHINA and WEHRHEIM 1996), and as penalties for failure to report to federal statistical bodies lost force, their reporting behavior is likely to color the objective indicators for more current years.

Macroeconomic distortions. During the first years of the transition period, various problems of macroeconomic instability existed, for example hyperinflation, exchange rate fluctuations and distorted capital markets. These problems were most severe in 1992 and 1993, though macroeconomic stability improved thereafter until 1997. However, at the end of 1997, contagious as well as spill-over effects of the world financial crisis stalled the recovery (SEROVA, VON BRAUN, and WEHRHEIM 1999: 351).

Taking these caveats into account, it has to be understood that the IOT used here is merely a starting point for an empirical version of an economy-wide model for Russia and therefore should be interpreted as a *second-best solution* only. All results obtained from simulations have to be interpreted against the background of this data base. Therefore, in the following we will discuss the specific steps which were mandatory to make it consistent with the model's needs and to update it.

The data base for the 1990 model

Balancing the 125–sector IOT for 1990

The 1990 IOT for Russia (WORLD BANK 1995) was computed in producer prices, contained 125 sectors, and, hence, was fairly well disaggregated. Additionally, it contained information on value added components (row-wise) and on final demand components (column-wise). It was structured in a similar way to the micro-SAM presented in Figure A3-2. Nevertheless, in order to compute the IOC for the Russia model in a consistent and useful way, two major manipulations of the original data in this IOT were mandatory:

- a *reduction in the number of sectors* to limit the information on sectors which are not the prime interest of this analysis;
- as the IOT *was not square* yet, and hence not fully consistent with SNA standards, the *IOT had to be balanced* both from the cost and the product approach.

Generally, a first step in the compilation of the data base for any economy-wide model is to tailor the sectoral disaggregation to the specific purpose of the study (see section 3.3). As the focus of this study is the agro-food sector, it was helpful that the original IOT provided information for six food industries and two agricultural production sectors – crop production and animal husbandry. However, when the first version of the Russia model was compiled, the seven food industries were kept separate, and agricultural production was totaled into one sector as the first model version was designed to address tax-policy issues in Russia from a macro-economic perspective rather than from the perspective of the agro-food sector (e.g. WEHRHEIM and WIEBELT 1997a). Other sectors such as manufacturing and services were aggregated into representative sectors in order to keep the model's data base at a manageable size. The 'light manufacturing' sector, for instance, contains a total of 13 sectors such as 'glass and porcelain', 'leather', and 'pharmaceuticals' (see Table A-3). Other important sectors in the Russian economy such as the 'fuel industry' were kept at a relatively low level of disaggregation (containing oil products, gas and coal production) so as not to hide the specific input-output characteristics typical for the raw material sectors. In contrast, the sector 'mechanical engineering' represents a highly aggregated sector containing about 30 sectors. No knowledge gain was expected from including further details on the components of this sector. Furthermore, it should be noted that some of the pre-transition characteristics are hidden in this aggregation scheme (see Table A3-3). For instance, in sector 16 ('trade & transport') of the 1990 IOT, not only freight costs have been aggregated but also trade mark-ups for which further information was available on neither their origin nor their structure. Sector 17 represents the 'service sector' containing almost all social services which were provided during the Soviet era to the population as public goods via the state (including health and education services). Additionally, this sector contains the defense sector, and the high level of secrecy about this sector meant that it was long associated with a number of question marks. Even though Steinberg had no specific information on this sector, he attempted to take it into account by representing at least some of its economy-wide dimension. For instance, he indicated the substantial material consumption of this sector, represented by a high IOC of the service sector from 'mechanical engineering', which obviously hides the expenditures for weapons produced for the army. Based on these arguments, a total of 17 sectors remained in the end, and these were used for the first model version.

In a second step, the 17–sector IOT for 1990 had to be balanced using both the cost and product approach, which turned out to be a non-trivial task given the available 1990 IOT. Even though it was compiled according to SNA standards, it contains some elements which were rather obscure. For instance, it shows various forms of wage payments: on top of 'wages', it also contained 'social security' payments, 'other wages' and 'bonus wages'. While the first two components were added to wage payments, basically yielding gross payments to workers, the third component was considered to be one form of capital remuneration and, hence, was added to capital costs.

After these components were allocated to the various macro-totals, another data manipulation had to be carried out to fully balance the IOT. The original IOT contained not only one import and one export sector but also two other additional trade sectors called 'net exports' on the final demand side and 'net foreign trade' as one cost component of domestic production, obviously used for balancing purposes. 'Net exports', however, had only two entries, one in the 'trade & transport' sector, which most likely represented trade mark-ups, and the other in 'services', which were earnings from exporting goods from the 'science' sector. While 'imports' were accounted for on the cost-side of domestic production, total 'exports' were given column-wise and, hence, as a final demand component.

One remaining problem with this IOT was the fact that three macro-totals were disaggregated as total sectoral costs components without further indicating the sectoral origin of these costs nor the sectoral origin of these components of final demand: turnover taxes, net subsidies, and imports. Hence, even though the whole table was implicitly balanced, it was not square in the sense that the calculation of sectoral domestic output and use was identical. In fact, total outlays and total revenues for intermediates were not identical. Information for the three positions was presented row-wise only, and, hence, the total costs of each sector and the total expenditures of each final demand component for these totals were given. What was lacking was the sectoral breakdown for each of these cost components in each sector as well as the sectoral disaggregation of the final demand value of the respective total. For each of these components, a separate input-output-matrix had to be calculated in order to obtain the total input-output value based on which the IOC were to be calculated.

A case in point are imports for which the original IOT indicated a total value of 144.891 billion rubles (bR), of which 68.856 bR were outlays of the economic sectors for imported intermediates. The row-wise representation of imports meant that only the information about the sector-specific *sum* of imported intermediates and the sum of imports bought by final demand components (households and investment demand) could be recovered from the IOT directly. The split of the remaining 76.035 bR between the final demand components could be retrieved from the 1990 IOT: 'households' spent 47.614 bR on imports and 28.421 bR were expenditures for imports by 'investment demand'. In order to fully balance the IOT, these final demand components had to be sectorally disaggregated. Therefore, the sector-specific column totals for imports (ΣM_j) were split up using the sector's weight in total expenditures for intermediates (a_{ij}). The respective value was then added to the absolute input-output value of the respective cell. The total expenditures of households and investment demand for imports were distributed among the sectors using the weights of households and investment demand for domestic products in order to trace the sectoral origin of imports for final demand. The same procedure was applied for the calculation of the matrix of net subsidies and turnover taxes. After adding these sector-specific values to the individual cells in the input-output table, the IOT was square and, hence, fully balanced.

The fully balanced IOT for 1990

Table A4-4 shows the IOC for all 17 sectors, the result of this first sector aggregation originating from the 1990 IOT. These IOCs were also used in a model used to carry out first simulations addressing the role of Russia's agro-food sector in the macro-economy (WEHRHEIM 2000a). Across columns, the table shows the sum of the IOC over all sectors and thus the total share of intermediates in each economic sector's total production costs. This IOT for Russia reveals some economic features which are characterizing the transitional status of the economy. For instance, specific manufacturing sectors such as the food industries rely heavily on intermediate consumption and, accordingly, produce only little value added. The share of intermediates used in the five food industries in this table ranges between 60% (sugar refining) and 94% (dairy production). The share of intermediate consumption of the agricultural sector is much lower. The lowest shares of intermediates used and, hence, the highest share of value added (exceeding 75%) was calculated for the two tertiary sectors, namely 'trade & transport' and the 'service sector'.

Updating the 1990 IOT to 1994

Between 1990 and 1994, the Russian economy underwent significant restructuring, induced by the first wave of macroeconomic reforms. Particularly the liberalization of prices and trade regimes as well as the privatization of firms and the concomitant downsizing of state involvement in economic activities were the driving factors behind some of the most significant changes in the Russian economy. Some of these changes were taken into account in an effort to update the IOT and the complete micro-SAM of the Russia model to reveal the situation in 1994:

- Updating macro-totals in the macro-SAM to 1994.
- Revealing the restructuring within the economy. Particularly the decline in manufacturing and the growth in the service sector.
- Disaggregating the agricultural sector to better reveal its dual structure.
- Adapting the domestic household demand and expenditure structure to more realistic levels.

These four steps in the updating procedure will be discussed in the following sections.

Updating macro-totals in the macro-SAM to 1994

The 1990 IOT has been updated to be consistent with 1994 macro data and consolidated government revenues and expenditures (IMF 1995). Table 4-2 shows the *fully consistent macroeconomic representation of the Russian economy in 1994* in trillion rubles (tR).[5]

The rules applied in the compilation of this macro-SAM were firstly to use officially released data for 1994 (from GOSKOMSTAT or IMF 1995) whenever available and secondly to use data from the 1990 IOT when such information for 1994 was lacking. Similar to the macro-SAM shown in Table A4-1, this macro-SAM specifies 8 macro accounts. While data for some cells was available from secondary statistics, other data had to be estimated in order to get a fully balanced macro-SAM.

For instance, because of a lack of information on value added figures, one crude assumption that had to be made was that the overall share of final demand in total domestic absorption remained constant between 1990 and 1994. The respective figures in the original IOT for 1990 (WORLD BANK 1995) reveal that the final demand share was 56.6%, yielding a share of intermediate inputs of 43.4%. Based on this assumption, and knowing that final demand totaled 628.2 trillion ruble (tR) in 1994, the amount of *total domestic absorption* can be calculated to equal 1110.0 tR in 1994. Hence, the total sum of *intermediates* was 498.8 tR. Furthermore, information from the IMF (1995) was available on the total sum of *labor remuneration* equaling 268.4 tR, corresponding to about 24% of total output or about 42% of GDP as calculated from this macro-SAM.[6]

---

[5] The average annual exchange rate of the ruble to the US Dollar was 2204 (OECD 1999a: 38).

[6] A similar share of GDP being spent on remuneration of labor is given in a more recent publication by GOSKOMSTAT (1999: 64) with 40.8% of GDP.

Table 4-2   Macro-SAM for the Russian Federation, 1994 (in producer prices; in tR)

	1 Activities	2 Commodities	3 Factors Labor	3 Factors Capital	4 Households	5 Government[a]	6 Capital acc.	7 Rest of World	Sum	
	1 ... n	1 ... n	1 ... m							
1	-	968.5	-	-	-	−7.3	-	152.4	1113.6	R
2	498.8	-	-	-	257.2	173.4	181.6	-	1111.0	E
3	-	-	-	-	-	-	-	-	-	V
Lab.	268.4	-	-	-	-	-	-	-	268.4	E
Cap.	343.3	-	-	-	-	-	-	-	343.3	N
4	-	-	250.9	61.3	-	-	-	-	312.2	U
5	3.1[b]	12.5	17.5	51.8[c]	14.5[d]	-	-	-	99.4	E
6	-	-	-	230.2[e]	40.5	−66.7	-	−22.4	181.6	S
7	-	130.0	-	-	-	-	-	-	130.0	
Sum[f]	1113.6	1111.0	268.4	343.3	312.2	99.4	181.6	130.0	-	
	E X P E N D I T U R E S									

Notes: a) The government account includes federal and regional budgets (consolidated). b) Indirect taxes: value added (43.4 tR) plus excise taxes (7.5 tR) minus subsidies (47.8 tR). c) Inclusive resource taxes (3.0 tR). d) Other revenues of the government. e) Retained profits and depreciation. f) Sums across columns indicate the payments or expenditures of each institutional account; sums across rows show the receipts or revenues.

Sources: IMF 1995; IfW 1998.

While the latter information was based on macro-data for 1995, another source gives an estimate of labor remuneration based on micro-data. KONINGS AND LEHMANN (2000) estimated the average share of wage payments as a share of total turnover of about 1000 'medium-sized and large' enterprises in 1996 at about 22%. This share of labor in total production costs, which is relatively high in comparison to industrialized countries (in Belgium, for instance, this share was about 12% in the mid–1990s), may be an indication of labor hoarding that is likely to be more severe in large enterprises often run on the grounds of soft budget constraints. Furthermore, LEHMANN, WADSWORTH and ACQUISTI (1999) pointed out that in spite of widespread incidence of wage arrears many firms in Russia still pay out wages to their workers in due course, which, in connection with declining revenues, may be another explanation for the high shares of wages in total turnover.

Capital absorption and indirect taxes were 343.3 tR and 3.1 tR, respectively (IMF 1995). The sum of these factor expenditures plus the sum spent on intermediate demand should equal gross output (column account) and also total receipts of the activity (row account: 1113.3 tR). At the same time, the data reveals that gross value added in the Russian economy amounted to 43.9% of *payments for labor*, while the remaining 56.1% was accounted for by *payments for capital factors*. This ratio is similar to the one revealed by an aggregate SAM for Russia for 1991 (BRABER and VAN TONGEREN 1996: 150), in which payments for labor made up 48.9% and payments for capital factors made up 51.9% of gross value added.[7] Furthermore, the macro-SAM for 1994 reveals that 17.9% of total capital income was distributed as *profits* to households and 15.1% as *taxes* to the government account. The capital account reflects the macroeconomic savings-investments identity. On top of *capital factor savings*, it receives *savings from households* (40.5 tR) while the negative *government savings* (66.7 tR) reflect the Russian budget deficit prevailing in the early transition period. It peaked in 1994 with more than 10.5% of GDP (IMF 1995: Table 21). Furthermore, the link between the capital account and the account for the rest of the world ensures that the second important macroeconomic identity, the balance of payments, holds. The 'negative revenues' of the capital account which are transferred to the rest of the world indicate a significant outflow of capital from Russia. These transfers amount to 22.4 tn rubles and could also be interpreted as capital flight. Government revenues stem from various sources: 3.1 tR are revenues from *indirect taxes*, 12.5 tR were *import tariff* revenues; *factor taxes* were 17.5 tR (on labor) and 51.8 (on capital) and *other household taxes* amounted to 14.5 tR (IMF 1995).

The preliminary estimate of GDP reported in WEHRHEIM and WIEBELT (1997a) and taken from the statistical appendix on the Russian Federation published by the IMF (1995: 18) was 630 tR. The revised GDP as calculated from the macro-SAM in Figure 4-1 amounts to 634.6 tR, as more recent data has become available in the meantime (IfW 1998).[8] *GDP at factor costs* incorporates capital costs (343.3 tR), labor remuneration (268.4 tR), indirect taxes (3.1 tR) and revenues from foreign trade (import tariffs and export taxes: 12.5 tR; 7.3 tR). It equals *GDP at market prices*, which is the sum of all final demand components including private consumption (257.2 tR), government expenditures (173.4 tR), the balance of trade (22.4 tR) and investment demand (181.6 tR). The latter reveals capital stock formation of which approximately 80% were investments and 20% stock-building. The slight variance of GDP taken from two different sources is due to higher revenues from external trade activities than originally reported.

---

[7] BRABER and VAN TONGEREN (1996) constructed a stylized CGE model for the Russian economy based on a data set for 1991 compiled from three different data sets to analyze energy price reforms in the Russian Federation. No information is given on the year to which the IOC refers.

[8] On the basis of the annual average exchange rate (ruble/US$ 2204), the Russian GDP in 1994 of 634.6 trillion rubles converts into 287.9 billion $ or an average per capita GDP of US$ 1942 compared to US$ 2405 in Poland or US$ 1759 in the Ukraine (calculated on the basis of data from WORLD BANK 1996).

The restructuring within the economy

One of the most typical features of the early transition period in all former centrally planned economies was the restructuring of the economy. In most former socialist countries, a dominance of heavy industries and a highly subsidized agricultural sector could be seen, which went along with inefficient resource allocation and use. Furthermore, the 'central planner', not the market, decided on the structure of the economy. The major share of gross output was produced in the state sector and in state-dominated firms. As soon as economic reforms released these firms into the market, and after the state had to withdraw its protecting hand from these firms, many of the firms and whole sectors of the economy proved to be no longer competitive under world market conditions and the given relative prices (see Figure 2-1 in chapter 2).

In Russia, the resulting restructuring of the economy was significant even in the early transition period and had a measurable impact on the sectoral composition of, for instance, total output and value added as early as the mid–1990s. While output declined during this period in most secondary sectors and in agriculture, the degree of the output decline in these sectors varies. Figure 4-2 reveals the trend in production for three industrial sectors and agriculture between 1990 and 1998. Since 1990, output in these sectors has decreased in comparison to the previous year, and the highest rate of output decline was experienced in 1994. Output contraction stabilized or even stopped in 1997 but due to the financial crisis continued in most sectors in 1998. The devaluation of the ruble following the crisis in mid–1998 opened a 'window of opportunities' by increasing the competitiveness of domestic industries (SEROVA, VON BRAUN and WEHRHEIM 1999). In fact, the degree of import substitution following the devaluation was greatest in those industries that produce consumer goods. This is particularly true for food products. While prior to the crisis the share of imported products in total food consumption in some major urban centers such as St. Petersburg and Moscow was higher than three quarters, it had dropped by the beginning of 2000 to about one third (ZMP Osteuropa, 10/2000).

The different trends in output decline shown in Figure 4-2 hint at some of the major trends in the restructuring of the Russian economy. A more detailed picture is revealed in Table 4-3 by showing the sectoral composition of both value added and total production. In the 1990 version of the Russia model, the sectoral composition of value added and total output was predetermined by the structure revealed in the IOT. For the 1994 version of the IOT, the share of the individual sectors in GDP as reported in GOSKOMSTAT (1998a) was used as a proxy for the new sectoral composition. The respective sectoral shifts between 1990 and 1994 are shown in Table 4-3.

The data highlights one of the obvious features of the restructuring process that took place in the first part of the transition period: the share of manufacturing sectors and agriculture declined both in terms of value added and total output. While the share of manufacturing sectors (sector 3 to 7 in Table 4-3) in total output fell from 33.8% to 29.9%, the share of agriculture in total output fell from 13.4% to 8.1%. In contrast, raw material-based industrial sectors, namely power and fuel industries, as well as the construction sector increased their share in total output and value added.

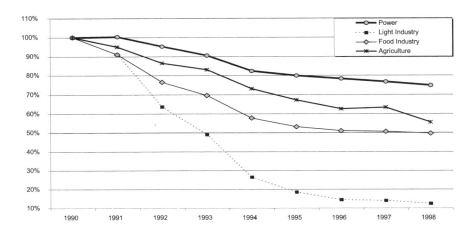

**Figure 4-2   Production indices for agriculture and industrial sectors
            (1990 = 100), 1990–98**

Sources: GOSKOMSTAT 1999; GOSKOMSTAT 1998a: 16.

The picture is slightly more mixed with respect to the food industries. While their share in value added fell in almost all cases between 1990 and 1994, the share of some sub-sectors of the food industry in gross national production increased in some cases. Despite the fact that the value added share of the flour and bread industry in total value added declined between 1990 and 1994, its share in total production increased. This might be due to the fact that bread products in particular have been highly subsidized by regional authorities (MELYUKHINA, QAIM and WEHRHEIM 1998). However, local production of bread was not only continued by many large-scale firms. Additionally, many newly founded small-scale bakeries were the first fully private enterprises in many urban centers. The fact that the bulk of bread products is rarely traded over long distances might have contributed to this slight increase in the relative importance of this sector.

While the share of the state-dominated service sector in value added and in total output remained relatively constant, the share of the trade sector with respect to both indicators increased. This hints at the increasing importance of trade activities in the course of transition (cf. KUHN 2000). The sectoral share in total investments declined in many of the secondary sectors, revealing their relatively low competitiveness as compared to primary sectors such as power and fuel industries, which attracted a higher share of investments. The decline in the share of total national investments was most significant in agriculture. On the eve of transition, in 1990, this sector received with 21.2% more than one fifth of total national investments. By 1994, this share had declined to just about 5%. It is evident that the capital stock in agriculture must have deteriorated significantly because of this reduction in investments.

Table 4-3 Sectoral composition of GDP and capital investments in the Russian Federation, in %

Sector	Share in value added		Total production		Allocation of national investments		
	1990[a]	1994[b]	1990[a]	1994[b]	1990[a]	1990[b]	1994[b]
1 Electric power	1.7	3.5	1.9	3.7	2.7	4.4	4.7
2 Fuel industry	3.8	9.5	3.7	7.9	6.1	11.6	13.0
3 Metal industry	3.4	2.7	5.2	5.7	4.7	2.9	3.6
4 Chemicals industry	2.4	2.0	3.5	3.0	3.4	1.7	1.5
5 Mechanical engineering	12.3	7.9	14.0	11.5	11.2	8.3	3.6
6 Wood industry	2.6	2.2	2.8	2.6	2.6	2.1	1.1
7 Light manufacturing	3.9	4.0	8.3	7.1	4.4	1.2	0.5
8 Construction	10.5	12.1	10.5	11.9	6.4	5.9	4.7
9 Sugar industry	0.2	0.1	0.3	0.3	0.3	n.a.	n.a.
10 Flour milling	0.4	0.4	1.5	1.8	0.5	n.a.	n.a.
11 Meat processing	0.6	0.5	1.7	1.7	0.8	n.a.	n.a.
12 Dairy processing	0.5	0.8	0.9	1.0	1.0	n.a.	n.a.
13 Other food industry	1.9	1.4	2.8	2.4	2.5	n.a.	n.a.
14 Animal feed	0.2	0.1	0.7	0.4	0.2	n.a.	n.a.
15 Agriculture[c]	15.8	8.5	13.4	8.1	21.2	15.9	5.0
16 Trade	13.0	17.8	9.0	12.9	14.6	29.4	37.2
17 Services	26.8	26.5	19.8	18.0	17.5	13.7	22.5
Sum	100.0	100.0	100.0	100.0	100.0	100.0	100.0

Notes: a) Calculations for 1990 derived from IOT published by WORLD BANK (1995). b) Calculations derived from GOSKOMSTAT 1998a. c) To enhance the comparability of the figures between both years, the total share of all agricultural sectors in 1994 is also given.

Disaggregation of the agricultural sector

In a next step, the empirical representation of the agricultural sector in the Russia model was adjusted to more realistically reveal its situation in 1994. In the 1990 version of our Russia model, one aggregate agricultural sector was shown (WEHRHEIM 2000a). In an attempt to increase the details of the analysis with respect to agriculture, we will distinguish four agricultural sectors in the 1994 model version. This is still much lower compared to other CGE country studies which have a focus on agriculture, but it is restricted by the limited availability of data for the various sub-sectors of Russian agriculture.[9]

To better address the issues of farm structures, these sectors have not been disaggregated by products but according to farm types. More specifically, the following four sectors are represented in the model: one crop and one livestock sector, each representing the *large agricultural enterprises* (LAEs) and, hence, the former collective farms. The LAEs were highly specialized during the era of central planning and in most cases continued to produce either livestock or crop products during the transition period. Additionally, we represent the two other types of agricultural producers that have become relevant in the transition period: the *household sector* and *privatized farms* (see chapter 1).

In a first step, the aggregate sector *agriculture* that was represented in the 1990 version of the model and for which the IOC were calculated with the data given in the 1990 IOT was split into these four sectors according to the shares of each sector in total agricultural output, as reported by GOSKOMSTAT (1998a) for 1994. While LAEs produced approximately 80% of total agricultural output in 1990, this share dropped to about 53% in 1994 (GOSKOMSTAT 1999). By 1994, the vast number of small-scale producers, mostly referred to as *household plots or private subsidiary plots* (Lichnie Podsobnie Khozyaistva/LPH), produced another major share of agricultural output (approximately 44%). The total number of such small-scale agricultural producers in Russia was in 1994 16.5 million (GOSKOMSTAT 1999: 217).

Additionally, most urban households maintain a private individual family garden on the outskirts of the cities, the well-known *datchas*, where substantial amounts of food are produced.[10] Nevertheless, GOSKOMSTAT does not take this production into account when calculating total agricultural output, as production from these private gardens is used purely for self-subsistence of urban families. According to official GOSKOMSTAT (1999: 217) estimates, the total number of such family gardens in 1994 was about 22.4 million. In fact, the number of households running small-scale plots varies significantly between regions in Russia and between the size of towns. In a household survey of three Russian oblasts (Orel, Pskov and Rostov), the share of households with access to a private garden in the capital of the region was much lower (77, 36 and 30%, respectively) than in other urban dwellings in the region (87, 58 and 54%). In rural areas, almost all households had either access to a household plot or a private garden in all three regions (99, 97 and 100%) (THO SEETH 1997: 140).

A third type of relevant agricultural producers in Russia emerged from reforms in the early transition period: the fully privatized and newly created *private farms*, 270,000 of which were operating mostly as family farms in 1994 (GOSKOMSTAT

---

[9]   WOBST (1998) presented a micro-SAM for Tanzania in which 21 out of 56 sectors belonged to agriculture. Such a high level of disaggregation is useful if the underlying data set reveals sufficient variation between the different sectors specified.

[10] A quantitative analysis of the impact of households' food production in rural and urban areas on subsistence, income, and poverty that is based on a household survey in three Russian regions is presented in THO SEETH (1997) and THO SEETH et al. (1998).

1999: 217).[11] In the IOT, its share in total output reached less than 1%. In fact, after 1994 the number of private farms stagnated (cf. SEROVA 2000), which meant that the officially reported share of private farms in gross agricultural output also remained at about the same level it had already reached in 1994 (3%).

Additionally, the data base of the model contains one sector that is obviously closely linked to agriculture: the *animal feed* sector is directly affiliated to the large agricultural enterprises. This sector was represented separately in the original 1990 IOT. Its input-output relations reveal its strong dependence on the two sectors representing the LAE but it also has typical links with the food industry. As a result of its high share of expenditures used for intermediates (more than 90.5%), it should be understood as an accounting artifice of the 1990 IOT and not as a primary agricultural production sector. To avoid coloring the data base of the two large agricultural enterprise sectors, it was not combined with these sectors.

In a second step the input-output relations of the four agricultural sectors had to be adapted to the situation in 1994. The underlying data is *synthetic* in that it originates from various primary and secondary data sources and was adapted to make it economically consistent. The input-output structures of the two sectors representing the large agricultural enterprises relate closely to those presented in the IOT for 1990. While the absolute value of expenditures for single intermediates was reduced in proportion to the decline in output of the LAE, the structure of the IOC remained more or less constant.

The specific procedure may be exemplified by looking at the four agricultural sectors' IOC, revealing the use of intermediates obtained from the *chemicals* industry. In the original 1990 input-output table, the IOC for agriculture's use of chemicals is 0.0243, which simply means that 2.43% of the sector's total production costs represent the purchase of inputs from the chemicals industry (see Table A3-4). The IOC for the four agricultural sectors in the IOT for 1994 reveal quite a different IOC for chemicals. The IOC for chemicals is highest in the LAE producing crops with 4.4% of the sector's total production costs. Chemicals are used to a much lesser extent by private subsidiary plots, private farmers and the LAE specializing in livestock production (2.3%, 0.5%, and 0.1% of the sector's production costs, respectively). EU farm network data (SCANDIZZO 2000) reveals not only much higher shares of chemical inputs in all crop sectors but also a higher IOC for chemicals, for instance, in vegetable production (0.20) as compared to cereal production (0.08). In the course of transition, input intensity has fallen considerably, which seems to indicate that such low levels of input use are justified. In spite of the low input intensity of household production in Russia, it is assumed that at least some chemicals are bought by this sector, as the major share of vegetables is actually produced by this sector (cf. VON BRAUN and QAIM 1999).

The LPH sector is not only typical for Russia but also for other CIS countries such as the Ukraine. KOESTER and STRIEWE (1999) argue that LPH producers in the

---

[11] The restructuring of Russia's farm sector has received much attention and is discussed in various publications. Background information on the early pattern of agricultural restructuring can be found, for instance, in BROOKS and LERMAN (1994) or WEHRHEIM (1997).

Ukraine are actually 'cross-subsidized' because they obtain industrial and on-farm inputs from the collective farms in substantial amounts. For Russia, this close link has been confirmed with case-study surveys of LPH producers in three Russian oblasts (Pskov, Orel and Rostov; SEROVA and KHRAMOVA 2000a). Official data from GOSKOMSTAT, which obviously does not take into account this cross-subsidization between large-scale farms and the LPH producers, indicates that the use of fertilizer is highest in the cereal sector which is to a great extent concentrated in the large agricultural enterprises. In fact, the close link between the *private subsidiary plots* and the *large agricultural enterprises* is one of the most distinct features of Russian agriculture in the transition period. A stylized representation has been used to indicate the input-output-relations between the four farm types, departing from the revealed input-output structure in the IOT for 1990: both sectors representing the LAEs make significant payments to the LPH sector. For crop-producing LAEs, these expenditures make up 6.6% of the sector's total production costs, while these transfers amount to almost 10% in the case of an LAE specialized in livestock production. The idea is to represent one obvious feature of Russia's rural economy in transition in the data base: while the LAE crop producers make payments in kind to their associated private subsidiary plots by leasing machines, transferring fertilizer and seeds, the LAE livestock producers are using feed as one major form of payment in kind to reimburse their workers for foregone cash income. These expenditures are basically a burden on the LAE and are likely to contribute to the high share of unprofitable LAEs.[12] Such expenditures are normally based on informal contracts between the two parties: the recipients are in many cases both workers and members of the LAE, many of which have 're-registered' as cooperatives.[13] While average wages for agricultural workers are among the lowest compared to all other sectors in the Russian economy, LAEs use *payments in kind* to their workers as compensatory payments. It is this interrelationship that is represented with the respective IOC.

There are a few other distinct features of the four agricultural sectors in the Russia model that have been taken into account in the compilation of the individual sectors' IOC and that are revealed in Table A4-4. In general, the share of intermediates is much higher in the two sectors representing the LAEs. This reveals the fact that these enterprises are more market-oriented, also with respect to inputs, at least when compared to the small-scale producers in the household sector and the private farms, which often suffer from insufficient access to input-markets due to various market imperfections (e.g. for Russia: TACIS 1998; for the Ukraine: PERROTTA 1999). At only 20%, the share of intermediates in total production costs is the lowest in the *LPH sector*. Additionally, about half of the intermediate inputs (10.9% of total production costs) used in this sector's production process stem from the sector itself. This high dependence on inputs produced within the same sector relates to the fact that this sector is to a considerable extent a *self-subsistence sector*

---

[12] According to different sources, between 60 and 80% of the large agricultural enterprises in Russia were unprofitable in 1998 (IANBYKH 2000; TILLACK and SCHULZE 2000; ZEDDIES 2000).

[13] For a discussion of the term 're-registration' see WEHRHEIM (1997).

and hardly relies on commercial inputs, for which the household plot owners would need to pay with cash they do not have. In contrast, there are at least some private farms that attempted early on in the transition period to improve their efficiency by buying inputs and new machines from the market. This is indicated with the respective values of the IOC of the private farm sector for inputs originating from the chemicals industry, light manufacturing, mechanical engineering sectors as well as from the power industry.

In addition to the distinct sectoral input relationships of these four agricultural production sectors, they also have rather different output structures revealed by the composition of sectoral revenues, as indicated by the row-wise comparison of sectoral IOC in the IOT shown in Table A3-4. Therefore, the forward and backward linkages for the four agricultural sectors indicates that the *degree of commercialization of the LPH sector is weakest* particularly compared with the two large scale agricultural sectors. The sugar industry receives 45% of its raw materials, in monetary terms, from the LAE crop sector. The meat processing industry buys substantial amounts of the raw materials it processes from the LAE livestock sector (amounting to 36% of its expenditures for intermediates). The share of inputs the meat processing industry is able to collect from the household plot sector makes up only 6% of the sector's total expenditures for intermediates, despite the fact that this sector represents more than 40% of total agricultural output in the model's data base and in fact produces substantial shares not only of crop but also of livestock products.

Summing up, the agricultural sector in the Russia model is represented by four types of producers, each of which has a characteristic production structure and/or typical levels of market orientation with respect to both inputs and outputs.

Up-dating household demand

Macroeconomic reforms in the transition process have had substantial micro-economic repercussions. Liberalization of prices and trade, for instance, resulted in a significant restructuring in the composition of household demand. Additionally, high inflation and declining real wages induced a restructuring of expenditure patterns. One of the most distinct features is the increase in the average share of total expenditures Russian households spend on food. The data in the IOT for 1990 shows a share of approximately 37% of total household expenditures spent on food. Data from the household survey carried out in three Russian oblasts in 1995 indicates that the average Russian household's share of food expenditure increased significantly in the transition period (Table 4-4). Taking into consideration market demand for food products only, the share of household expenditures for food averaged about 56%. If subsistence production is taken into consideration, this share increases to almost 80%. Official Russian statistics indicate for 1995 (1996) a share of 49% (47%) of total expenditures spent on food (GOSKOMSTAT 1998a).

Therefore, household demand had to be updated to more realistically represent households' behavior in the model. The structure of private consumption was adjusted to the revealed expenditure shares for *market demand only* as they were obtained from the household survey (Table 4-4). However, in some cases this data can not be considered to be fully representative for Russia as a whole.

The two large urban centers, Moscow and St. Petersburg, were not taken into account in the survey, which led to the decision to set the overall food expenditures share in the 1994 IOT slightly lower and, in turn, to increase the share of expenditures for some goods from the manufacturing sectors.

The mechanical engineering sector comprises, for instance, cars that enjoyed high revealed preferences particularly in Moscow and St. Petersburg in the early years of transition. Another case in which the data from the household survey and macroeconomic information did not match is the flour and bread industry. The household survey indicated that a share of 10.2% of an average household's expenditures was spent on flour and bakery products. Using this high consumption share in the IOT, however, the total amount of officially reported production and imports together could not satisfy this high demand. To balance the sector's expenditures and receipts, the household expenditure share of this commodity group was reduced to 8.7%. As can be seen from Table 4-4, the share of expenditures for food in the 1994 IOT is significantly higher as compared to the respective share in the IOT for 1990 (51.1% versus 36.8%). It still seems to be rather low if it is compared to the respective expenditure share that has been calculated from the household survey. However, it has to be borne in mind that it is very difficult to take subsistence production and, hence, one part of the virtual economy with all the associated difficulties of evaluating non-monetary transactions, fully into account in the model's data base.

As in the case of the flour industry, using an even higher share of household expenditures than the one calculated for the 1994 IOT would yield significant imbalances within the whole model.[14] Household demand for non-food sectors would then be too low to balance the overall supply in these sectors (e.g. light manufacturing).

A second major feature of household expenditures in the 1994 IOT as compared to the one in 1990 is the high share of food stemming from the food industry. This is due to the fact that the respective shares shown in Table 4-4 also comprise the expenditures for imported food products. Due to high revealed preferences for imported food brands, the share of imported processed food products rose significantly during the transition period, while that for imported agricultural bulk commodities, particularly of grains, decreased.

---

[14] One alternative would be to differentiate between different socio-economic household types which reveal different levels of food expenditure shares. This aspect has been neglected as this study focuses on allocative instead of distributional issues.

Table 4-4 **Sectoral composition of household expenditures in the Russian Federation, in % for 1990, 1994, and 1995**

Sector	House-hold expend. shares 1990 IOT[a]	House-hold expend. shares 1994 IOT[b]	Household expend. shares from surveys[c] in 1995	
			with self-production	market demand only
1 Electric power	0.9	3.4	1.7	3.5
2 Fuel industry	0.5	1.3	0.8	1.6
3 Metal industry	0.1	0.2	0	0
4 Chemicals industry	1.2	3.1	1.5	3.1
5 Mechanical engineering	5.8	1.5	0.1	0.2
6 Wood industry	2.6	1.0	0	0
7 Light manufacturing	14.3	14.2	7.5	15.3
8 Construction	0.1	3.6	0.8	1.6
9 Sugar industry	0.9	4.2	2.3	4.7
10 Flour milling	2.8	8.7	5.0	10.2
11 Meat processing	6.5	10.0	3.4	7.1
12 Dairy processing	3.7	7.0	3.9	8.0
13 Other food industry	7.3	8.4	3.8	7.8
Food industries	21.1	38.3	18.4	37.8
Total industry without construction	53.5	63.0	30.8	61.5
14 Animal feed	0	0.1	0	0
Agriculture	15.7	13.5	59.9	17.8
15 LAE crop	6.9	0.2	n.a.	n.a.
16 LAE livestock	8.8	2.5	n.a.	n.a.
17 Private subsidiary plots (LPH)	n.a.	10.5	n.a.	n.a.
18 Private farms	n.a.	0.3	n.a.	n.a.
Agriculture and food industries	36.8	51.1	78.3	55.6
19 Trade & Transport	16.4	5.6	2.2	4.5
20 Services	21.4	14.2	7.1	14.5
Sum	100	100	100	100

Sources: a) Derived from the IOT for 1990 (WORLD BANK 1995). b) Data from GOSKOMSTAT 1997a. c) Shares have been computed with household data (calculated with data from surveys carried out in cooperation with the Center for Economic Analysis, Moscow, and the Institute for Food Economics and Consumption Studies, University of Kiel; the survey is fully documented in: THO SEETH (1997) and derived from the IOT for 1990 (WORLD BANK 1995).

The balanced micro-SAM for 1994

After the four updating steps mentioned above, it was mandatory to balance the complete micro-SAM again. The technical procedure was split into three steps. First, the input structure was kept constant. Only the composition of macroeconomic features and the row and column sums of sectors and institutions were adjusted to reveal the predetermined structure as indicated by the updated macro-SAM. Secondly, the input-output section of the micro-SAM had to be balanced. In doing so, the sectoral revenues from selling inputs and the expenditures for buying intermediates were adapted to reveal the macro-structure as shown in the macro-SAM and the new sectoral structure in the economy. The major assumption in this procedure was to keep value added shares constant. Therefore, the column totals were adapted proportionally, while the row sums, and hence the revenues from selling inputs, had to be adapted to obtain a balanced micro-SAM. Thirdly, the sector-specific difference between row and column totals that arises was distributed between the cells of the respective rows and columns by minimizing the total variation of the input-output values in each sector.

The result of the above steps is a *fully balanced micro-SAM*, which consists firstly of a new sectoral distribution of all macro-totals in the macro-SAM and secondly of a new set of IOCs (see Table A4-5). The new sectoral distribution of the most important macro-totals has been summarized in Tables 4-5 to 4-8. Table 4-5 gives an overview of the sectoral distribution of the macro-totals which, when aggregated, yield the total value of *sectoral domestic production calculated from the demand side*. The sums across columns for each of these macro-totals correspond to those given in the macro-SAM (see Table A4-2). Table 4-6 expresses the sectoral value of the macro-totals as a share of total domestic production in each sector and the share of sectoral imports in exports. Tables 4-7 and 4-8 are set up accordingly, but they reveal the *sectoral distribution of macro-totals by calculating total domestic production for each sector using the production approach*. Again, the column sums have to correspond to those given in the macro-SAM. Additionally, the row sums of Table 4-5 have to be equal to the row sums in Table 4-7. Table 4-8 again reveals the relative share of macro-totals in the sectoral value of domestic production.

Demand for intermediates and private consumption. A significant share of total domestic absorption in Russia is used for intermediates (45%) (see Table 4-5). Total household consumption accounts for about 23% of domestic absorption. The sectoral share of revenues obtained from selling the good produced as an intermediate to other sectors varies significantly. The share of intermediates in total use of domestic production is lowest in the construction sector (13%), which sells only 7% of its domestically produced 'good' to private households. At 23% of total revenues, it is also low in the meat industry, while household demand for the good from the meat industry is 33% higher than total use of domestic production, indicating that Russia's significant import demand for meat is due to the deteriorating efficiency in the meat-processing industry and in the livestock-producing large-scale agricultural enterprises. All of the heavy industries and primary product sectors deliver the major share of their domestic production to other sectors as inputs, while private consumption accounts for only smaller shares. In contrast, the sectoral share of private consumption in domestic production of light industry and all food industries is dominant, again in some cases exceeding domestic production due to imports. The sectoral share of the farm sector's domestic production that is used for private consumption is lowest for the LAE crop sector, which produces bulk commodities normally not ready for consumption. This share is higher for the LAE livestock sector, as meat can be used by private households without the need for further processing as is the case for grains. The highest private consumption share among the four farm sectors is registered for the LPH producers (69%), indicating the strong links between this sector and household consumption.

Government demand and gross investments. The transition from plan to market coincided with a significant decline in the overall importance of the state. Hence, government demand generally was scaled down in the course of transition. The same holds true for gross investments, as the state was no longer able to finance investments at the level common during the pre-transition period. On the one hand, the share of government demand in domestic production – at 16% – is relatively low compared to other former socialist countries. A share of about 16% of domestic production being used for gross investments is, on the other hand, relatively high compared to other countries in transition, and it almost reaches the levels in western economies. A second feature of government demand in the course of transition is a substantial diversification in contrast to the pre-transition period, induced by the privatization of former state firms. Nevertheless, the major share of total government demand was satisfied by the two service sectors.

Table 4-5　Sectoral distribution of macro-totals for the Russian Federation in 1994; calculation of total use of domestic production (UDP) using the consumption approach, in bR

		Reve-nues for inter-mediates	Private con-sump-tion	Govern-ment con-sump-tion	Invest-ment demand	Domes-tic absorp-tion	Imports after customs	Domes-tic absorp-tion – imports	Exports – export taxes	Use of domes-tic pro-duction
		1	2	3	4	5=1+2+3+4	6	7=5–6	8	9=7+8
1	Power	27717	8802	410	0	36930	675	36255	5079	41333
2	Fuel	30324	3365	1481	0	35170	6555	28615	58766	87380
3	Metallurgy	29813	514	1290	13806	45424	6982	38442	25393	63834
4	Chemicals	19299	7973	1956	1816	31044	16200	14844	19008	33852
5	Machines	58832	3858	23135	53220	139046	27822	111223	16687	127910
6	Wood	19754	2470	1043	1888	25155	2280	22875	5804	28679
7	Light	56916	36423	1794	785	95918	20376	75542	3482	79025
8	Construction	17767	9259	4571	104278	135876	3250	132626	0	132626
9	Sugar	1495	10929	198	0	12622	9925	2697	84	2782
10	Flour & bread	7854	22390	195	0	30439	10200	20239	216	20455
11	Meat prod.	4431	25753	1290	0	31474	12380	19094	232	19326
12	Dairy	3142	18034	492	0	21667	10580	11087	185	11272
13	Other food	9421	21556	312	0	31289	8218	23072	3750	26822
14	Animal feed	4390	61	78	0	4529	0	4529	0	4529
15	LAE:[1] crops	21458	518	3617	0	25593	2150	23443	461	23904
16	LAE:[1] livest.	15667	6521	3162	0	25350	2840	22510	392	22902
17	LPH[1]	11889	26913	0	0	38802	0	38802	0	38802
18	Private farms	2262	692	0	0	2954	0	2954	0	2954
19	Trade & trans.	96759	14279	26768	2449	140255	1967	138288	5462	143750
20	Service	59610	36890	101605	3358	201463	0	201363	0	201463
	Macro-totals[2]	**498800**	**257200**	**173400**	**181600**	**1111000**	**142500**	**968500**	**145100**	**1113600**

Notes:　1) LAE: Large Agricultural Enterprise; LPH: Private subsidiary plots. 2) Macro-totals correspond to respective figures in the macro-SAM (see Table 4-2).
Source:　Based on balanced micro-SAM for 1994.

Table 4-6   Sectoral shares of macro-totals in total use of domestic production (UDP) and sectoral ratio of imports to exports for the Russian Federation for 1994, in %

		Reve-nues for interme-diates	Private demand	Govern-ment demand	Invest-ment demand	Imports	Exports	Imports/Exports
		in % of domestic production						*in %*
1	Power	67.1	21.3	1.0	0.0	1.6	12.3	13.3
2	Fuel	34.7	3.9	1.7	0.0	7.5	67.3	11.2
3	Metallurgy	46.7	0.8	2.0	21.6	10.9	39.8	27.5
4	Chemicals	57.0	23.6	5.8	5.4	47.9	56.1	85.2
5	Machines	46.0	3.0	18.1	41.6	21.8	13.0	166.7
6	Wood	68.9	8.6	3.6	6.6	7.9	20.2	39.3
7	Light	72.0	46.1	2.3	1.0	25.8	4.4	585.1
8	Construction	13.4	7.0	3.4	78.6	2.5	0	0
9	Sugar	53.7	392.9	7.1	0	356.8	3.0	11766.6
10	Flour & bread	38.4	109.5	1.0	0	49.9	1.1	4719.1
11	Meat prod.	22.9	133.3	6.7	0	64.1	1.2	5337.2
12	Dairy	27.9	160.0	4.4	0	93.9	1.6	5734.0
13	Other food	35.1	80.4	1.2	0	30.6	14.0	219.1
14	Animal feed	96.9	1.3	1.7	0	0	0	0
15	LAE:[1] crops	89.8	2.2	15.1	0	9.0	1.9	466.4
16	LAE:[1] livest.	68.4	28.5	13.8	0	12.4	1.7	724.5
17	LPH[1]	30.6	69.4	0	0	0	0	0
18	Private farms	76.6	23.4	0	0	0	0	0
19	Trade & trans.	67.3	9.9	18.6	1.7	1.4	3.9	35.4
20	Service	29.6	18.3	50.4	1.7	0	0	0

Notes: 1) LAE: Large Agricultural Enterprise; LPH: Private subsidiary plots. 2) Macro-totals correspond to respective figures in the macro-SAM (see Table 4-2).
Source: Based on balanced micro-SAM for 1994.

Government demand for services accounts for more than 50% of total domestic production in the *service* sector but only for about one fifth of total domestic production in the trade and transports sector. This indicates the far-reaching dominance of the state remaining in the service sector. Particularly in rural areas, most social services such as health and education are still provided as public goods exclusively by the state (LOHLEIN and WEHRHEIM 2000). The second most important sector for government demand is *machinery*, revealing the high demand for military equipment.

In contrast, the *structure of investment demand* by sectors of origin has not been adapted significantly. Hence, it is assumed that the major share of investment demand is still met by the construction sector (57%) and the machine industry (29%).

An interesting feature of *government demand* is the fact that the two sectors representing the large agricultural enterprises together are the third most important source of government demand (together about 6%). This is due to the fact that one important instrument of agricultural policies in the transition period was *government food stocks*. While deliveries to these stocks were obligatory for LAEs in the pre-reform period, by 1994 they were *de jure* non-obligatory. Hence, it is no surprise that the quantitative relevance of these stocks had declined by 1994. However, federal and regional stocks were built up to cover the needs of the military, the 11 million inhabitants in the northern areas of Russia and, at least to some extent, of the two largest urban centers in Russia, Moscow and St. Petersburg as well as to ensure the food supplies of a region to guarantee its food security (MELYUKHINA and WEHRHEIM 1996: 15 f.). In 1994, deliveries to federal and regional stocks comprised, for instance, 46% of all 'officially marketed' grain and 76% of all beef and poultry (GOSKOMSTAT 1999). The term 'officially marketed' grain, however, refers to the volumes marketed by the former collective farms only. Even though GOSKOMSTAT claims to take barter transaction of cereals into account, widespread incidence of underreporting is likely.

The actual share of gross agricultural output collected by regional and federal stocks is likely to have been much smaller in 1994. Incentives to merchandize parts of their production volume via these stocks were in many cases provided by regional governments, which made access to either subsidized inputs or credit for large agricultural farms dependent on deliveries to these stocks (SEROVA and KHRAMOVA 2000b). These stocks explain the relatively high share of government demand in total domestic production of the crop and livestock producing LAEs. To a much lesser extent, these government stocks also bought food from the food processing industry. It is a typical feature of agricultural policy in Russia that neither private farmers nor private subsidiary plots were able to sell their products via these stocks to the government in exchange for intermediates. This is an additional feature in the data base, indicating the different economic and political situation of the LPH and private farm sector in Russia's transition period.

Imports.   In the course of transition, the composition of Russia's imports changed significantly, which is also true for the structure of agricultural imports. Before transition started, agricultural imports consisted mainly of bulk commodities. Early on in the 1990s, the revealed preferences of Russians for western food brands increased tremendously, resulting in a sharp increase in imports of consumer goods. At the same time, the lack of subsidies for the LAE livestock sector and its deteriorating competitiveness resulted in a sharp reduction in imports of feed. For

1994, the IMF statistic estimated the share of food imports in non-CIS trade to be 29%. As the Russian Federation also imported food from other CIS countries (e.g. the Ukraine), the share of food imports in total imports in our micro-SAM has been estimated to be higher (39%). The high *revealed preferences of Russian consumers for imported, mostly processed food* items is indicated by the generally high shares of imports in domestic production in all food industry sectors. The sector with the highest share of imports in the micro-SAM for 1994 is the sugar industry, indicating the low competitiveness of this sector in Russia. Non-food sectors with considerable shares of imports are some of the manufacturing sectors (chemicals, light industry and mechanical engineering), while the resource-based sectors such as the power and fuel sector reveal rather low import shares.

There are four sectors that do not demand any imports at all, namely, the two small-scale farm sectors (LPH and private farms), the animal feed sector, and the service sectors. These sectors produce exclusively for the domestic market and do not export any share of sectoral production. The construction service imports at least a minor share (2.5% of use of domestic production), revealing the trend towards relying on foreign construction firms, at least with respect to major infrastructure projects in the major urban centers of the country.

Exports. In contrast to most other CIS countries, Russia is a resource-rich country. According to IMF data (1995), Russia's exports consisted mainly of two commodity groups of products in 1994: energy (gas, oil), minerals, and metals made up the major share of Russia's exports (42 and 27% respectively). Again, the IMF figures refer to non-CIS trade only, while our data has to incorporate trade with CIS countries as well. As there are high energy exports to other former members of the Soviet Union which are resource-poor, for example Moldavia, we report an even higher share of energy exports (44%). These energy-based exports prevented Russia from running into a trade deficit in the course of the transition, as imports of consumer items, and particularly of food products, increased significantly. Instead, in 1994, Russia realized a trade surplus of 22.4 trillion ruble. While the revenues from raw material based exports reduced the pressure to speed up reforms, the fact that the country's exports consist of only a few commodities makes the economy vulnerable to external shocks. Indeed, the financial crisis of 1998 was induced to some extent by the openness of Russia's energy sector, which was exposed to significant reductions in world market prices – a classic case of a *spill-over effect* (SEROVA, VON BRAUN and WEHRHEIM 1999: 351). Conversely, the increase in world prices for energy products in the course of 1999 helped to reduce the government budget and the encourage recovery of the economy.

The next three important export categories in 1994 were chemicals (particularly fertilizer), timber (raw and processed), and machinery (IMF 1995). Most of these exports were taxed in 1994, enabling the Russian government to collect 7.3 tR of revenues. Exports of agricultural and food commodities were – at least in 1994 – not taxed, with the exception of fishery products. The IMF (1995) statistic for Russia indicated that this category made up 3.3% of Russia's total export value in 1994.

The inclusion of the fishery sector in the food industry sector 'other food' therefore also explains most of this sector's exports and export taxes in the 1994 micro-SAM. The same four sectors in the Russia model which did not import anything did not export any share of their sectoral output either. The composition of imports and exports in the micro-SAM highlights the fact *that intra-industry trade* is not only theoretically feasible (see chapter 3) but also empirically a common feature in most sectors. However, this is obviously dependent on the degree of aggregation.

Table 4-7   Sectoral distribution of macro-totals for the Russian Federation for 1994; calculation of total origin domestic production (ODP), using the production approach, in bR

		Expend -ures for inter- mediates	Indirect taxes	Capital (Deprec. & Profits)	Wages	Total value added	Origin of domes- tic pro- duction
		1	2	3	4	5=2+3+4	6=1+5
1	Power	19855	367	17750	3362	21479	41333
2	Fuel	28974	898	49295	8213	58406	87380
3	Metallurgy	47232	185	12473	3945	16603	63834
4	Chemicals	21556	431	6761	4105	12297	32852
5	Machines	79128	1275	27931	19575	48781	127910
6	Wood	15215	412	7325	5727	13464	28679
7	Light	54175	2182	14170	8498	24850	79025
8	Construction	57966	367	27103	47190	74660	132626
9	Sugar	1951	10	655	166	831	2782
10	Flour & bread	18316	136	1233	771	2140	20455
11	Meat processing	16146	220	2433	526	3179	19326
12	Dairy	6583	32	3519	1138	4689	11272
13	Other food	17944	352	6466	2059	8878	26822
14	Animal feed	4099	6	305	119	430	4529
15	LAE:[1] crops	11317	−2472	7847	7212	12587	23904
16	LAE:[1] livestock	14953	−3134	3563	7519	7948	22902
17	LPH[1]	9143	0	17956	11703	29659	38802
18	Private farms	1172	3	1090	689	1782	2954
19	Trade & transport	34407	853	70711	37779	109343	143750
20	Service	38669	977	63713	98104	162794	201463
	Macro-totals[2]	498800	3100	343300	268400	614800	1113600

Notes: 1) LAE: Large Agricultural Enterprise; LPH: Private subsidiary plots. 2) Macro-totals correspond to respective figures in the macro-SAM (see Table 4-2).
Source: Based on balanced micro-SAM for 1994.

Table 4-8 **Sectoral shares of macro-totals in total origin of domestic production (ODP) and ratio of imports to exports in each sector for the Russian Federation for 1994, in %**

		Interme-diates	Indirect taxes	Capital	Wages	Value added
		EI/ODP	*IT/ODP*	*CAP/ ODP*	*WAG/ ODP*	*VAD/ ODP*
1	Power	48.0	0.9	42.9	8.1	52.0
2	Fuel	33.2	1.0	56.4	9.4	66.8
3	Metallurgy	74.0	0.3	19.5	6.2	26.0
4	Chemicals	63.7	1.3	22.9	12.1	36.3
5	Machines	61.9	1.0	21.8	15.3	38.1
6	Wood	53.1	1.4	25.5	20.0	46.9
7	Light	68.6	2.8	17.9	10.8	31.4
8	Construction	43.7	0.3	20.4	35.6	56.3
9	Sugar	70.1	0.4	23.6	6.0	29.9
10	Flour & bread	89.5	0.7	6.0	3.8	10.5
11	Meat processing	83.5	1.1	12.6	2.7	16.5
12	Dairy	58.4	0.3	31.2	10.1	41.6
13	Other food	66.9	1.3	24.1	7.7	33.1
14	Animal feed	90.5	0.1	6.7	2.6	9.5
15	LAE:[1] crops	47.3	−10.3	32.8	30.2	52.7
16	LAE:[1] livestock	65.3	−13.7	15.6	32.8	34.7
17	LPH[1]	23.6	0	46.3	30.1	76.4
18	Private farms	39.7	0.1	36.9	23.3	60.3
19	Trade & transport	23.9	0.6	49.2	26.3	76.1
20	Service	19.2	0.5	31.6	48.7	80.8

Notes: 1) LAE: Large Agricultural Enterprise; LPH: Private subsidiary plots.
Source: Based on balanced micro-SAM for 1994.

The sectoral share of exports in the total use of domestic production is highest in the fuel industry (67%), followed by the chemicals industry, indicating that Russia is an important exporter of fertilizer. The sectoral share of exports in domestic production was also high in the metal industry and in another resource-based sector, the wood industry, which is of particular relevance for the northern and eastern regions of Russia.[15]

However, the relatively *high number of sectors* that are either completely *isolated from world markets* or only engage in one-way trade is a typical feature of an economy that is still on its way from plan to market and, hence, in transition. In fact, it will be one of the necessities to improve the linkages between these sectors with world markets in order to enhance the transmission of price signals onto domestic markets.

After having discussed the sectoral structure of domestic production using the consumption approach, we will now turn to the *composition of the origin of domestic production* (Table 4-7 and 4-8). The absolute amount (498.8 tR) and the share of *expenditures for intermediates* in total domestic production (45%) have to be identical to the revenues obtained from sales of intermediates and the respective share in the use of domestic production. The other expenses of the various economic sectors are value added components (indirect taxes, capital, and wages).

Expenditures for intermediates. Expenditures for intermediates account for the major share in sectoral production costs for 14 out of the 20 sectors in the Russian economy. They make up less than 50% of total sectoral expenses only in the resource-based fuel industry (33%), the construction sector (44%), the trade and the service sector (24 and 19%) and the two small-scale agricultural production sectors (24 and 40%). While the respective share for the four sectors mentioned first make intuitive sense, the low share of intermediates in the two small-scale agricultural production sectors is at first glance surprising. This is, however, due to the fact that these sectors reveal a *low degree of commercialization* that is also revealed in a low degree of input use.

While *indirect taxes* make up a minor share of total domestic production, the data in the respective column in Table 4-7 reveals one other feature that has been carried over from the original IOT (WORLD BANK 1995): indirect tax payments of some

---

[15] In fact, the structure of Russia's economy varies significantly between regions. In Siberia, where only about 25% of the country's population lives, approximately 75% of the country's raw material production is located. In contrast, regional production of food is concentrated in the European part of the country and the south-western part of Siberia. It would obviously be highly relevant to take such differences into account in a 'model economy', as they also coincide with a different importance of transaction and trade costs as well as with different political policy regimes. It is therefore also likely that, for instance, different trade policies, and hence integration strategies, will have regionally different welfare and efficiency implications. Since these issues are of pivotal importance for a country of the spatial dimension of Russia, they have been addressed in a complementary study within the same research project. The description of the model and policy simulations with the respective bi-regional, stylized, computable general equilibrium model for Russia, based on data for 1995, can be found in KUHN (2000).

sectors are negative, indicating that these sectors are subsidized. It is no surprise that these subsidies were allocated almost exclusively to farm sectors in the 1990 IOT. In fact, only the two farm sectors representing the large-scale enterprises received such subsidies in 1994.[16]

In addition to indirect taxes, *capital and wages* are the two major value added components in all sectors. The sum of value added in each sector is (as a share of 100%) the mirror image of expenditures for intermediates. The share of wage payments is in most sectors lower than the share of capital in total sectoral production costs. Exceptions are three sectors in which production is very labor-intensive: the construction sector, services, and the LAE livestock sector. In the three sectors power, fuel, and metallurgy, wage costs make up less than 10% of total production costs. The same holds for the food processing industries. While all four agricultural sectors are relatively labor-intensive, the share of labor in total production costs is highest in the two LAE sectors. In contrast, the capital intensity is highest in the LPH sector, referring to the high untaxed and implicit profits being realized because of low opportunity costs of labor.

Support to agriculture, import tariffs, and export duties

Support to agriculture. For 1994, the OECD (1999a: 231) estimated the aggregate support to Russian agriculture to be tR –2.607, indicating a taxation of agriculture. However, this estimate was caused by an implicit tax Russian agriculture had to pay due to lower domestic prices received by producers as compared to world market prices for comparable agricultural commodities. The respective negative transfer to agriculture that was due to this negative market support component amounted to tR –13.598. It was only partially offset by three forms of direct support granted to Russian agriculture via government programs: tR 1.6 for direct payments; tR 4.0 for reduction of input costs; tR 2.8 for general services (OECD 1998: 231).[17]

The sum of direct payments and reduction of input costs (tR 5.6) granted to Russian agriculture is reflected, as indicated in the previous section and in the micro-SAM, by negative indirect tax payments amounting to 5.6 tR which are exclusively channeled to the two agricultural sectors representing the large-scale

---

[16] In the IOT for 1990, the major share of these negative indirect taxes and, hence, subsidies was allocated to livestock production, which means that the share of subsidies granted to this sector is even higher (MELYUKHINA and WEHRHEIM 1996). In spite of low direct federal subsidies for crop-producing agricultural enterprises, it is assumed here that the agricultural subsidies are evenly split between the former collective farms, constituting one form of soft budget constraint that is common practice with respect to many of these large-scale farms. In fact, credit subsidies have been the major form of state subsidies for Russian agriculture during the transition period. Today, the restructuring of debt arrears of the large-scale agricultural enterprises is one of the major issues of Russian agricultural policy. Because this model does not have an independent investment module that is linked with a money market, these effects can only be represented indirectly.

[17] A detailed discussion of the various forms of policy instruments used to support agriculture in 1994 and a quantification of the support levels granted to agricultural producers in various regions can be found in MELYUKHINA and WEHRHEIM (1996).

farms producing crops and livestock. This sum of direct transfers to the collective farm sector in the model is equal to a share of 0.8% of GDP or about 3% of total government expenditures. About one fifth of the 'reduction of input costs' consisted of subsidies for inputs such as fertilizers, while the other four fifths of this support component were credit subsidies that were granted to both livestock- and crop-producing LAEs. Out of the direct payments to Russian farmers in 1994, the largest share was livestock subsidies. Therefore, in our model, the share of these direct support components received by LAE livestock producers is – at 56% – higher than those received by LAEs specialized in crop production.

Private farms and agricultural production on private subsidiary plots does not receive any state support in the model. Despite the fact that a major share of gross agricultural output today is produced on private subsidiary plots, agricultural policy makers in Russia tend to understate the relevance of household production for income generation and food consumption (VON BRAUN and QAIM 1999). Therefore, agricultural policy in Russia resists enhancement of the production possibilities of this sector with state interventions. However, even though these two sectors are widely exempted from paying taxes, the private farm sector has to pay indirect taxes. Instead of direct support, we indicate the close relationship between the large-scale and the small-scale farms via an input relationship between this sector and the large-scale farms, thereby hinting at the above-mentioned 'cross-subsidization'. While we had to express this input relationship in monetary amounts, in reality it is a barter transaction in which the owners of small-scale farms barter their labor for variable inputs from the large-scale farms such as feed, sugar, or other products.

Import tariffs.   Up to 1993, imports of agricultural commodities were actually subsidized to meet the domestic need for food. These subsidies were abolished after July 1993 and import tariffs were levied on most agricultural commodities from March 1994 on. According to the OECD (1998: 123), import tariffs range from 0% to 25%. Specific tariff levels reported were: 0% for vegetable oil, 1% e.g. for wheat and sunflower, 8% for beef and pork, 10–15% for dairy products, 20% for poultry, and 25% for potatoes. Hence, there is a moderate degree of tariff escalation in two ways: the tariffs on raw materials are, at least for plant products, generally lower than for processed products. Secondly, the variance of tariff rates is significant, but on this product level not as high as the customs code for all food commodities would reveal. If we look at the three- or four-digit level of product aggregation, tariff rates for some food items were over 50% as early as 1995 (e.g. for wines and spirits).

Since the sectoral disaggregation in our model is different, these import tariffs had to be adapted: all industrial sectors in the model have import tariffs of approximately 10%. Exceptions are the chemicals and light industries, which have import tariffs of 15%. The import tariffs of the food industries in our model resemble the structure as reported by the OECD for 1994 as closely as possible. Specifically, we chose the following import tariffs: 6% for the sugar industry,[18] 1% for the flour industry, 15% for meat processing, 10% for the dairy sector, and 5% for

---

[18]   In the model, this is basically a weighted import tariff for the high rate for white sugar and the lower rate for raw sugar cane imported from Cuba.

the other food industry. The large agricultural enterprises producing crops are protected by low import tariffs of only 1%, while the large livestock producers are protected at a higher level with an import tariff of 10%.

Export duties. After a legislative reform of the customs code that became effective after October 1st 1993, no export duties were applied to major agricultural exports. A 10–25% export duty on grains was foreseen by the legislation but was never applied. Hence, we did not introduce any export duties for agricultural exports in our model. Instead, the Russian government's revenues from export taxes in 1994 were collected by taxing raw material exports. In our model, we apply a flat rate of 5% to non-agricultural exports.

Calibration of parameters and solution of the Russia model

Calibration of parameters

A final step in developing the data base of an applied CGE model is the calibration of the model's parameters. In general, there are two options in obtaining the parameters for a CGE model. First, *parameters can be estimated empirically* from time-series or cross-sectional data. The other possibility is to 'calibrate' *the model parameters* by matching a consistent data set for the base period (the micro-SAM) with a fully consistent theoretical model (MANSUR and WHALLEY 1984). The latter refers to a theoretical structure that consists of various macroeconomic identities and which ensures that the *Walras' law* holds (see chapter 3). The choice of method depends on the objective of the studies. If the quality of parameters has a high priority, an econometric estimation would be most appropriate. If a higher priority is given to disaggregation and the flexibility of the model's data base with respect to changes in the data base, the second method is commonly used. This is so because the mandatory data for empirical estimates of parameters is normally not available. Particularly for transition countries, time-series estimations of relevant parameters are scarce and particularly in the case of Russia almost non-existent. This is due to the structural break that occurred when transition from plan to market began, and which does not yet permit the use of annual data.

In order to maintain the consistency between the data base and the theoretical model, the number of calibrated model parameters may not exceed the number of independent relations in the theoretical structure of the model, implying that the choice of functional forms has to be kept relatively simple. This is one of the reasons why most CGEs for developing and transition economies rely on simple functional forms such as Cobb-Douglas production functions and Constant-Elasticity-of-Substitution functions on the demand side. In fact. this also highlights the general trade-offs and conflicts between theoretical complexity, transparency of the model, and the extent to which the model is able to replicate reality.

Due to the high level of disaggregation chosen in the 1994 version of the model (20 sectors) and the particular problems associated with any econometric estimations of parameters in the transition period, the parameters used in the Russia model were calibrated. For this procedure, elasticities of substitution of import demand and elasticities of transformation of export supply had to be predetermined.

These elasticities are 'synthetic' in that they have been collected from the literature and were not estimated empirically (see chapter 4).

The *calibration procedure* can be explained as follows: at first, all transactions in product (D_i, M_i, Q_i, E_i, X_i, V_i, C_i, G_i, I_i) and factor markets (F_i) are given in value terms only. To separate the respective information, all domestic prices for goods (P_i^x, P_i^d, P_i^k, P_i^q, P_i^e, P_i^m) and all factor prices (W_f) are set equal to 1, a procedure commonly used if the focus of interest is on relative price changes only. As a result, the value and quantity of the respective variables are identical. Given the import tariff rates (t_i^m) and export subsidy rates (t_i^e) and a given exchange rate (R), the export and import prices of each sectoral good denominated in foreign currency ($P_i^{\$m}$ and $P_i^{\$e}$) can be calculated. Given the input-output coefficients (a_{ij}) that have been calculated from the micro-SAM (see Table A4-2), and given indirect tax rates (t_i^x), the net prices (P_i^v) can be calculated.

To further specify the supply side of the Russia model, the parameters of the Cobb-Douglas-function and those of the transformation function between domestic use and exports are needed. Substitution between labor and capital is specified using an elasticity of substitution of 1. This implies that a percentage change in the relative prices of factors results in an equal change of opposite sign in the use of factors in the production process. Hence, the factor shares in the production process of each sector remain constant, whereas the share of the factor in total sectoral output is equal to its production elasticity. As the factor shares are known from the micro-SAM, the production elasticities can be calculated from the factor demand equation in which the numerator indicates the total factor payments of sector i and the denumerator the level of gross value added in the sector:

$$\alpha_i^f = \frac{W_f \cdot \varphi_{if} \cdot F_{if}}{P_i^v \cdot X_i} \tag{4-1}$$

Once these production elasticities have been calculated, the parameter (a_i^x) indicating the sector-specific efficiency of production can be calibrated using the following formula. This parameter is relevant, as it can be exogenously altered in single sectors in simulations, implying shifts in the sector-specific total factor productivity. It can be calculated based on the following formula:

$$a_i^x = \frac{X_i}{\prod_f F_{if}^{\alpha_{if}}} \tag{4-2}$$

Furthermore, due to the fact that the quantities and hence the values of sectoral exports (E_i) and imports (M_i) are given, the sectoral distribution parameters which are needed to specify equation 3–20 can be calibrated (see chapter 3.). The respective formula is obtained by transforming the equation for maximizing revenues:

$$\delta_i^t = \left[1 + \frac{P_i^d}{P_i^e} \cdot \left(\frac{E_i}{D_i} \right)^{1/\sigma_i^t} \right]^{-1} \tag{4-3}$$

Since the sectoral output values (X_i) in the base year are also given, and if the elasticity of transformation is exogenously determined, the equation can now be reformulated for the supply of goods in such a way that the value of the scale parameter a_i^t can be calculated:

$$a_i^t = \frac{X_i}{\left[\delta_i^t \cdot E^{\rho_i^t} + (1 - \delta_i^t) \cdot D^{\rho_i^t}\right]^{1/\delta_i^t}} \tag{4-4}$$

In quite a similar way, the parameters on the import side can be calibrated if the sector-specific values for the elasticity of substitution are known. As all prices are set equal to one in the base period $P_i^d = P_i^m$, the values for D_i and M_i can be used directly from the SAM. Rearranging the equation for import demand, the distribution parameter can be calculated with the following formula:

$$\delta_i^q = \left[1 + \frac{P_i^m}{P_i^d} \cdot \left(\frac{M_i}{D_i}\right)^{1/\sigma_i^{qt}}\right]^{-1} \tag{4-5}$$

If the absolute value of the elasticity of substitution is known, we now can also calculate the scale parameter a_i^q:

$$a_i^q = \frac{Q_i}{\left[\delta_i^q \cdot M^{\rho_i^q} + (1 - \delta_i^q) \cdot D^{-\rho_i^q}\right]^{1/\rho_i^q}} \tag{4-6}$$

Hence, in order to calibrate all of the above-mentioned parameters, the values of the exogenously determined trade parameters must be specified.

Exogenously determined trade parameters

Elasticity of substitution and transformation. As we allow for two-way trade, we have to specify the degree of substitutability between domestically and internationally traded products. On the demand side, the CES determines to what extent the imported good can be substituted by the domestic good stemming from the same sector and vice versa. Hence, the CES indicates the ease at which imports and the domestic good are substituted for one another, given a change in the relative price between the domestic and the imported commodity. On the supply side, the CET indicates the degree of homogeneity between the sectoral good that is sold domestically and the good from the same sector that is exported. In a way similar to the CES, the CET determines the flexibility with which a production sector can shift its resources to produce more for either the domestic or the export market in response to a change in relative prices between the domestic and the exported commodity.

Several factors affect the absolute value of the CES and CET. First, as mentioned in chapter 3, the degree of price transmission between domestic and world markets is dependent on the share of imports and exports in respective domestic markets (DE MELO and ROBINSON 1989). Second, and closely linked to the previous factor, the degree of product aggregation in the respective sector has an effect on the elasticity values. A rule of thumb is that the value of both elasticities is expected to increase with the level of product disaggregation and vice versa. Hence, higher CET and CES would be expected for sectors which produce rather homogenous products such as raw materials. Third, the choice of these elasticities is related to the time frame considered. As our model is defined to reveal short-run effects of policies, elasticities should not be set too high.

The matter is complicated because *empirical estimates of trade elasticities are rather scarce* and are not available from the literature for Russia. The structural break experienced by transition economies has made empirical estimations of such elasticities that are based on times series impossible, at least in the early transition period. Therefore, the elasticities in the Russia model are 'synthetic' to the extent that they have been chosen based on CGE models for other middle income countries (see Table 4-9).

Elasticity of transformation (σ_i^t). FAINI (1988) estimated a long-run elasticity of transformation of 2.9 for a bundle of consumer goods for Turkey. WIEBELT (1996: 130) used this value as a benchmark for the manufacturing sectors in a one-country CGE model for Malaysia. As our model is specified to analyze short-run developments, this value will define the upper bound for CET values. WEYERBROCK (1996) specified CETs for the aggregate of the FSU in a multi-country CGE model with a focus on agriculture. The CET values in her model are also synthetic in as much as they are based on values reported in the literature. The CETs applied by WEYERBROCK (1996: 29) for the FSU varied significantly. The underlying assumption for the choice of CET values was that firms in net exporting sectors can be more responsive to changes in the relative prices for the domestic good and the exportable. Hence, the highest CET values were assigned to the energy sector (CET=2.9), while CETs for manufacturing sectors were set around 2 and for food and agricultural commodities between 1.5 (for meat) and 0.75 (for cereals and other grains). In a CGE model specified for Hungary, BANSE (1997) used rather uniform CET values. Out of nine sectors, the CET for the two service sectors was set at 0.6, while the CET for all remaining sectors was set at 2 – including the two agro-food sectors (agriculture and food industry).

Based on the studies discussed above and on an evaluation of the above-mentioned factors against the background of the Russian economy, the CET values in the Russia model range between 0.5 and 2.9. The latter value was used for Russia's fuel sector, which is very competitive in international markets and which produced about 40% of the country's total exports in 1994. A value of 2.0 was chosen for the other major exporting sectors that produce raw materials or semi-processed products. A CET value of 1.5 was chosen for the mechanical engineering sector and light manufacturing. A rather low CET value of 0.75 was chosen for the trade sector, despite its relatively large export share, as this sector produces a relatively non-homogenous product group.

Table 4-9   Trade elasticities in the Russia model

Sector	Constant elasticity of transformation (CET) between domestic sales and exports σ_i^t	Price elasticity of export demand η_i^t	Constant elasticity of substitution (CES) between imports and domestic goods σ_i^q
1 Power	2.0	−3	0.6
2 Fuel	2.9	−3	0.6
3 Metallurgy	2.0	−3	0.6
4 Chemicals	2.0	−3	0.6
5 Machines	1.5	−3	1.5
6 Wood	2.0	−3	0.6
7 Light	1.5	−10	2.0
8 Construction	−	−10	−
9 Sugar	1.1	−10	1.5
10 Flour & bread	0.75	−10	1.5
11 Meat prod.	0.5	−10	1.5
12 Dairy	0.5	−10	1.5
13 Other food	1.1	−10	1.5
14 Animal feed	−	−	−
15 LAE:[1] crops	1.1	−6	1.5
16 LAE:[1] livestock	0.5	−6	1.5
17 LPH[1]	−	−	−
18 Private farms	−	−	−
19 Trade & trans.	0.75	−10	0.5
20 Service	−	−	−

Sources: Elasticities are 'synthetic' and not empirically estimated. They are derived from other comparable country models for Hungary (BANSE 1997), for the FSU (and other countries; WEYERBROCK (1998); Malaysia (WIEBELT 1996) and are adapted to the specific economic circumstances in the Russian Federation in 1994.

For the *food industries, lower CET values were chosen* for the following reasons. Surveys of food processing firms in various regions in the European part of Russia (1995 and 1998) and in Siberia (in 1998) indicated that most food processing firms had significant problems exporting to foreign markets, even if relative prices became more favorable (SEROVA and KHRAMOVA 2000a). Responsiveness to higher world market prices in the middle of the 1990s was higher for some selected agricultural raw materials, particularly cereals which were exported at least in small quantities even at times when Russia had to import a significant amounts of cereals. Therefore, crop products and products from the sector "other food" as well as sugar were typified as – relatively speaking – more homogenous goods with a CET value of 1.1. A lower CET value was chosen for meat and livestock products, for which concerns about food quality and food safety are additional factors, which means that domestic production can not easily be transformed into international supply.

Export demand elasticity.   The export demand elasticity is inversely related to a country's share in world markets. At the same time, the export demand elasticity can be taken as an indicator for the degree of homogeneity between a country's exports and other countries' exports. The choice of the export demand elasticity therefore depends on both characteristics of the respective sector.

Five sectors in the Russia model do not export or import anything at all (see Table 4-5) and therefore produce non-tradable goods only. Hence, no export demand elasticity has to be specified for these sectors. The exports from all other sectors of the Russian economy compete in world markets with respective products of different origin. Energy products and other raw material exports such as oil, gas, or wood are relatively homogenous, which meant that infinitely elastic export demand could be assumed for these products. However, Russia is certainly not a 'small country' in these markets, so a downward sloping export demand curve is more realistic. For the food industries, the respective elasticity has been set at a value of $\eta_i^t = -10$ while for the LAE sectors an elasticity of –6 was chosen.

Elasticity of substitution.   The above-mentioned studies are also indicative for the choice of CES values. WIEBELT (1996: 137) set the CES value for agricultural sectors, forestry, and fisheries at relatively high levels of 4.0, that for mining at 3. Lower CES values were used for light industry and the food industry (1.5), while the lowest values were set for industrial intermediates (0.66) and the trade and service sectors (0.5). The CES values in the above-mentioned CGE model for Hungary varied between 1.5 and 0.4. The higher value was used for the food industry, for agriculture, and the chemicals sector, while the lower CES value was used for the two service sectors (BANSE 1996: 108). In the multi-country CGE model, WEYERBROCK (1998: 29) set the CES for the FSU at a higher level: here, the lowest value was 0.95 while the upper bound of sectoral CES was 4.0. The CES values in the non-food sectors were set at the lower end of this scale and the upper end was reserved for primary agricultural products. CES values for crop products were higher than those for animal products.

CES values in the Russia model.   Generally, the share of imports in Russia's agro-food sectors was higher than in the non-agricultural sectors in 1994. Furthermore, these sectors were net importing sectors. Therefore, the CES values for the agro-food sectors are generally higher than the CET values. In the raw material producing sectors, low CES values of 0.6 were chosen. The only two industrial sectors which received higher values were the mechanical engineering sector (1.5), where domestic products were substituted for imported ones early on in the transition period. The same trend was observable in light industry, where imported products flooded Russian markets right after the trade restrictions were dismantled (CES=2.0). In the food industries, the CES values were generally set at 1.5. A CES value of below one (0.5) was chosen only in the trade sector.

Solution of the model

The theoretical model as it was described in chapter 3 consists of a system of interdependent, in some cases non-linear, functions that has to be combined with the data set discussed in this chapter. Due to the updating procedure of the micro-SAM from 1990 to 1994, we were in a position to compile two model versions: the *1990 version of the model* refers to the 17–sector model based on the original IOT as reported by the World Bank, while the *1994 version of the model* is based on the micro-SAM for 1994 which was described in detail in the above sections.

The theoretical and the empirical structure of the model have been programmed in the GAMS software, commonly used for the solution of applied CGE models. This is advantageous as GAMS offers also various algorithms (solvers) with which the equilibrium values of the endogenous variables can be determined. While the non-linear equation in older CGE models of the ORANI type were mostly linearized, a procedure that dates back to JOHANSEN (1960), GAMS solves the non-linear equations directly and simultaneously. Furthermore, GAMS has the advantage of offering a code that resembles normal algebraic notation whenever applicable.

Before we start with the discussion of policy simulations, it is advisable to look at the *solution procedure* with which GAMS searches for equilibrium values of all endogenous variables. Even though the system of equations is solved simultaneously, various parts of the solution mechanism can be identified. If the prices of all goods and all production factors (W_m) as well as the exchange rate have been determined, the sectoral factor prices (W_{mj}) can be determined as they linearly depend on W_m. Then the sector-specific factor demand, which depends on the marginal value product, can be estimated. In a next step, the model tests if, given the factor and goods prices, factor demand and supply are in equilibrium. If this is not the case, the factor prices are adjusted in such a way that a balance is reached. Once the use of production factors has been determined, the gross and net production in each sector can be determined based on the production function and by using the input-output coefficients. If the sectoral net production is evaluated at the given prices, the result is gross domestic product that is reallocated to the institutional accounts in the economy – predominantly to households as factor enumeration. In a next step, the sectoral quantity of supply is compared with the respective demand. If equilibrium is not reached, the price vector is altered stepwise to gradually reach a steady state.

Calibration of the model. While many model parameters such as input-output coefficients and various elasticities are exogenously determined, others are not (see chapter 4). The values of these missing parameters can be obtained by forcing the model to find an equilibrium which complies with all theoretical requirements specified in the theoretical version of the model such as market clearance, utility maximization on the consumer side and profit maximizing behavior on the production side. Furthermore, the base run solution has to replicate exactly the specified and balanced data base for the base year. The model solution that meets both of these requirements will entail absolute values for the missing parameters. This procedure to estimate the missing parameters from the empirically balanced and theoretically consistent base run model is called *calibration*. Based on these starting values, these parameters can also be changed in simulations.

The complex structure of the theoretical and empirical part of the model means that inconsistencies can not be excluded *a priori* when the model is calibrated and solved. Hence, it seems to be a requirement to carry out various tests with such models. In general, four types of 'tests' on the model's robustness can be carried out:

- test for homogeneity,
- additional consistency checks,
- sensitivity tests, and
- validation of the model.

A *test for homogeneity* can be a simple but powerful device, as it quickly reveals whether the theoretical and empirical parts of the model are consistent. This test is a first simulation with the model in which the central price index, in our case the GDP deflator, is – *ceteris paribus* – altered exogenously. Here, it was raised by 10%. The underlying concept of homogeneity implies that all real variables are homogenous to the degree 0 with respect to price changes. For instance, real GDP or sectoral output should not be altered in such a simulation, while nominal variables should increase by exactly 10%. Additionally, it has to be expected that all endogenous prices are changing to the same extent, while relative prices do not change. Indeed, the solution of this experiment yielded changes in all prices and nominal variables by 10%. All real and all exogenously determined variables (depending on closure rules, for instance factor supply, total government demand etc.) remained constant.

Additional consistency checks.　　One of the most essential checks for theoretical and empirical consistency of the model is to ensure that the *model is square*. *Theoretical squareness* refers to *Walras' law* which, as discussed in chapter 3, stipulates that the number of equations in the model is exactly the same as the number of variables (see Table A3-1 and A3-2). For this purpose, a summary table with model statistics can be installed at the end of each model run, providing the information on both the number of equations and the number of variables. In our model, the number of equations and variables was – at 439 – the same, and hence theoretical consistency was maintained. Furthermore, not only the test for homogeneity but all solution files of experiments should replicate the *empirical base run solution* as it has been predetermined in the data base of the model. Here, it is most important to program the model in such a way that it also calculates a macro-SAM similar to the one in Table 4-2 for the base period and one based on the new solution at the end of each program run. The model would be empirically square if the row account and the column account of each actor are equal in both cases. The macro-SAM for the base period should also yield the same value for macroeconomic variables as it has been specified in Table 4-2. Together, these tests ensure the theoretical and empirical consistency of the model.

One additional check can be made, but depends on the closure rule. As all domestic prices for goods (P_i^x, P_i^d, P_i^k, P_i^q, P_i^e, P_i^m) and all factor prices (W_f) are set equal to 1 in the base period, the output file which calculates a value for the base and the solution scenario for each endogenous variable should show a value of exactly 1.000 and 1.100 for all sectoral prices in the base period and the solution scenario, respectively.

Sensitivity tests are commonly carried out to test whether the model is sensitive to changes in the base run of the model. Therefore, sensitivity tests can generally consist of two different alterations: either changes in the data base or changes in the economic features of the model. We will restrict our sensitivity tests to two commonly used options: changing trade parameters and changes in the closure rules that determine which are decisive for the kind of equilibrium that can be identified. The purpose of these tests is to investigate whether marginal changes in simulation results can be traced back to changes in the model economy and if these changes are intuitively correct and theoretically plausible. The sensitivity tests with the trade elasticities are particularly important because the chosen values in this model are not based on empirical estimates.

Finally, *validity* tests would be based on the question as to what extent the model can actually replicate the economy-wide effects of real world developments ex post. The underlying idea is to ask whether the model – which is abstract by its very nature – is able to replicate economic developments that were observed in the past as a result of real world policy events or external shocks.

Chapter 5

Economic and Policy Simulations with the Russia Model

In the previous two chapters, we outlined the economic and empirical components of the economy-wide model for the Russian economy. In the next section, we will use this computable general equilibrium model for various economic and policy experiments. We will start out with a critical assessment of the Russia model and then provide an overview of the simulations. The simulations will look into the effects of domestic economic and policy changes, as well as changes in trade-related matters. While most of the simulations will be based on the fully updated model version for 1994, some simulation results are based on the 1990 version of the Russia model. This '1990 model version' refers to the Russia model which used the input-output-coefficients calculated from an input-output table for 1990, consisted of 17 sectors, and which has been described in WEHRHEIM (2000a). The '1994 model' is an update of this first model version in that household expenditures, input-output coefficients and other data sets have been updated to reveal the Russian economy in the year 1994 as closely as possible. The Russian economy in this model version is also disaggregated further and comprises a total of 20 sectors (see chapter 4).

Critical assessment of the Russia model

Some remarks of caution have to be made when interpreting the simulation results obtained from the Russia model. One typical feature of models in general is that they abstract from reality. In fact, just as a map on a 1:1 scale would be of no use to a car driver who wants to drive across Russia, a model of the country's economy on a 1:1 scale would be senseless for any policy analyst. Abstraction from details is therefore a requirement to understand the major forces driving an economy. However, we can not claim that we have left out only unimportant details. Due to the high level of in-transparency involved in the process of economic restructuring in Russia, we had to concentrate on some of the most characteristic features of the country's economic situation in the middle of the 1990s. Other features of economic reality in Russia, such as corruption, and/or the 'virtual economy' which is related with phenomenons such as wage and payment arrears, barter trade, and debt defaults, are far too complex to be captured with a computable general equilibrium model. In fact, the comparative-static model presented here presumes a 'closed system' instead of an 'open system' being driven by economic agents who learn and alter their behavior. Hence, it is a clear example of a positivist policy model for looking into quantitative if-then relationships. As SCHMID and THOMPSON (1999) stated, such a positivist approach may neglect the 'deep' causes which drive economic systems.[1] In fact, a better understanding of the 'deep' causes of the transition process in Russia's agro-food sector would necessitate a far broader perspective by taking institutional changes and non-economic factors such as social norms, cultural traditions, path dependence, and even psychological factors in explaining human behavior into account. Nevertheless, the complex economy-wide mechanisms that are set in motion by specific economic events have to be understood better. In particular, if the economic environment in which the economic development of Russia's agro-food sector took place in the first transition decade is better understood, it should become easier to identify the most appropriate policy options to enhance the sector's perspectives. Hence, what is true for models in general is of particular relevance to this economy-wide model for Russia: it can only highlight some aspects of the economic situation in the Russian economy; it can not claim to be holistic.

An obvious area for criticism of any CGE model for a transition economy is the data base. Due to the multitude of variables and the high degree of insecurity associated with many empirical values of these variables, it could be tempting to criticize individual data components and then come to the conclusion that if the data is wrong the model results are wrong as well. However, the model results are always interpreted against the background of this specific data set. In fact, it has to be understood that the results are strongly affected by the data specifying the benchmark. For example, a policy experiment that grants export subsidies for a sector with few exports might stimulate sizable *relative* export growth, which, from an economy-wide perspective, might be marginal in *absolute* terms. Hence, different economic constellations in the reference period have the potential to alter the economic effects of policy experiments. We will therefore use the discussion of simulation results to enhance the transparency of the mechanisms which drive the macroeconomic and the sectoral results. If these mechanisms are made transparent, the interested reader will be in a better position to evaluate similar events in the Russian context, even though the reference period might have changed significantly in the meantime. In order to indicate the impact of changes in the data base, we will contrast the results of some simulations carried out using the 1990 version of the model with those carried out using the 1994 version of the model.

Other limitations of CGE models are also obvious: even though the degree of disaggregation chosen in our model is relatively high, it is of course important to bear in mind that the individual sectors represented in the model do not produce homogeneous products as the model suggests. However, a low degree of product differentiation is a general problem of any general equilibrium or economy-wide model. Therefore, partial equilibrium models, which assume that general equilibrium effects are exogenous and constant, can be an important alternative because of their potential to include more product-specific market and policy characteristics (cf. FOCK et al.). The combination of partial and general equilibrium modules in one modeling

---

[1]   SCHMID and THOMPSON (1999: 1164) argue that attention to 'deep causes' would shift the focus of economists from merely changing events within given structures (such as the Russian economy) to changing event possibilities by transforming institutions.

framework has the potential to combine the comparative advantages of both approaches. While BANSE and TANGERMANN (1998) used a sequential approach by combining partial and general equilibrium effects, WOBST (2000) developed a CGE model that specifies alternative sectoral trade regimes within the CGE model. An attractive feature of this approach is the possibility to contrast the notion of perfect substitutability normally applied in partial equilibrium models with the notion of imperfect substitutability and, hence, the Armington assumption.

The incorporation of such features would be useful extensions to our current model. Therefore, it is obvious that the model in its current form is still stylized with respect to various characteristics of the Russian economy. Nevertheless, it can be used to show important general equilibrium responses to economic changes which occurred in the transition process. Taking the strengths and weaknesses of the model into account, we will use it in the following to trace some of the important economy-wide general equilibrium effects of exogenous shocks which have affected the agro-food sector in the first transition decade.

Overview of simulations

In chapter 2 we briefly discussed some of the most important developments of Russian agriculture in the early transition period. The strong output decline experienced by this sector was the subject of intensive domestic debate. Many explanations centered around two arguments: first, the worsening of agricultural terms of trade due to a more rapid increase in agricultural input prices in relation to agricultural output prices; second, the inflow of food imports was made responsible for the decline of domestic agriculture. Other sector-specific factors such as the deterioration of the capital stock or simply the reduction of total factor productivity due to the changes in the behavior of economic agents, as well as sector-neutral factors, received relatively little attention.

It is impossible to address all factors which have been suggested in the course of this debate within a comparative static CGE framework. Therefore, in the following, we will address some of these factors and discuss their linkages between the agro-food sectors and the rest of the economy.

Table 5-1 provides an overview of the major experiments which were simulated. We conducted five sets of experiments, each of which consists of three single experiments. Some of these 15 experiments were also simulated under different closure rules and after having changed the predetermined elasticity values. The results of the simulations will be reported for each set of experiments with two types of tables: one will show the changes of macroeconomic indicators (e.g. exchange rate, GDP, exports and imports) and percentage changes of sectoral output aggregated over various sectors; the other table will report the changes of specific indicators (change in output, output price, value added price, etc.) for individual sectors.

As mentioned above, the simulations are built around two dichotomies of factors which affected agricultural sector development in the transition period and which are likely to drive it in the future (see chapter 2 and figure 5-1). On the one hand, the dichotomy between *sector-specific* and *sector-neutral* factors (cf. MUNDLAK 2000) is relevant. On the other hand, the dichotomy between *domestic and international factors* which affect the development of agriculture will be essential. Furthermore, as discussed in chapter 2, the relevant economic effects that were set into motion in the transition process can be separated into a *price and policy effect* and *a technology effect*.

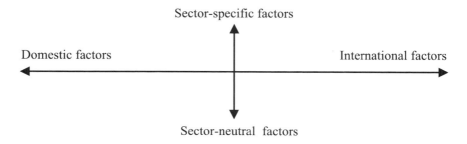

Figure 5-1   Two dichotomies of factors affecting agricultural development in the transition period

In a *first set of simulations*, we deal with factors that have been blamed for the poor performance of Russia's agro-food sector in the 1990s. One of the experiments looks into the effects of deteriorating terms of trade in agriculture due to increases in import prices for industrial inputs, a deterioration of agriculture's capital stock, and changes in the TFP in agriculture. First, results of these experiments that were obtained with the '1990 model version' will be reported.[2]

The *second set of simulations* replicates the same three experiments, but with the model version that has been updated to 1994. All other simulations will be carried out with the 1994 version of the model described in detail in chapters 3 and 4. SEDIK (2000: 183) pointed out that it would be interesting to see a comparison of results using the 1990 model and more recent IOC. In fact, one of the major differences between the two model versions refers to a specific comment of SEDIK (2000: 182) – the 1994 version of the model reveals much more realistically the structural diversity in agricultural production which has emerged during the first years of the transition process. Therefore, the simulations with the 1994 model should be helpful in showing the different effects of specific economic shocks on the various farm types which have emerged in Russia in the first few years of the transition period and which represent different production technologies.

---

[2]   Results obtained from simulations with the older model version have been reported in various previous publications, including e.g. WEHRHEIM 2000a; WEHRHEIM and WIEBELT 1998.

Table 5-1   Overview of simulations

Topic of simulations	Table[c]	Experiment 1	Experiment 2	Experiment 3
Set 1: Agriculture's output decline: sector-specific factors (1990 model version)	5-2	Increase in import prices in some sectors (declining terms of trade)	Deterioration of capital stock	*Increase in* TFP in large-scale agriculture[a]
Set 2: Agriculture's output decline: sector-specific factors (1994 model version)	5-4	Increase in import prices in some sectors (declining terms of trade)	Deterioration of capital stock	*Increase in* TFP in large-scale agriculture
Set 3: Agriculture's output decline: transport issues[b]	5-6	Reduction of TFP in transport sector	Red. of TFP in transport and reduction in domestic support for LAEs[a]	Increase in import prices in all sectors (declining terms of trade)
Set 4: WTO: Increasing agricultural protection	5-10	Increase in import tariffs in all agro-food sectors	Export subsidies for agro-food sectors	Increase in direct domestic support for large-scale agric.
Set 5: WTO: Liberalization of Russia's trade regime	5-12	Removal of agric. import tariffs and domestic support	Removal of import tariffs in all sectors of the economy	Introducing a uniform tariff rate (flat rate)

Notes: a) TFP = Total Factor Productivity; LAE = Large Agricultural Enterprises. b) All experiments from Set 3 onwards were simulated with the 1994 version of the model only. c) The number of the Table which reports the major macroeconomic changes is given in the second row of this column.

The *third set of simulations* looks into the consequences of a deteriorating infrastructure in Russia's transport sector and its effects on agro-food sectors. With the *fourth set of simulations*, we will discuss Russia's current strategy of integrating its agricultural sector into the WTO, which is currently a matter of intensive debate. Due to the relatively moderate level of protection granted to agriculture in Russia in 1994, which is also reflected in the data base, we turn the normal liberalization experiment around. Instead of analyzing the effects of liberalizing agriculture, we ask in this section what would happen if Russia were to gradually increase preferential treatment of agriculture by using either higher import tariffs, by introducing export subsidies, or by increasing direct domestic support to large-scale agricultural producers. With *the fifth set of simulations*, we will look into the effects of contrary trade strategies, i.e. what would happen if Russia were to further liberalize its trade regime under the pressure of WTO accession.

Explaining the poor output performance of Russian agriculture in the early transition process

Before discussing the simulation results, it is essential to recall how we have specified the model economy: our *Keynesian version of the model* is a short-run model, characterized by the following standard closure (see chapter 3). With perfectly elastic supply of labor at given nominal wages, employment is determined endogenously by labor demand. Real government spending and real investment have been fixed at their initial levels. Finally, the trade balance has been kept constant, reflecting Russia's limited access to foreign capital. The exchange rate of the ruble has been determined by financial markets, which means that the exchange rate in our model adjusts to reach external equilibrium. Exports and imports adjust in accordance with the relative price change induced either directly by the exogenous shock or the indirect changes due to the realignment of the exchange rate.

Experiment design for sector specific factors

The following three simulations were carried out with the 1990 version of the model (Set 1) as well as with the 1994 version of the model (Set 2). The objective of these experiments is to highlight some of the causes for the poor performance of Russia's agro-food sector in the 1990s and their relative importance, as well as to highlight the economy-wide effects of a potential recovery of the sector. Experiments 1 and 2 of Set 1 and 2 directly investigate the reasons for the poor performance of the agricultural sector in the 1990s. In Experiment 1, the prices for three important industrial intermediates are increased to address the issue of changes in relative prices and a deterioration in agriculture's terms of trade. In fact, this is the most prominent argument put forward in explaining the poor performance of the agro-food sector in the transition process. The second experiment analyzes the impact of a deteriorating capital stock in the agricultural sector and its impact on agricultural output. In contrast to the previous two experiments, the third experiment assumes a positive attitude and looks into the macroeconomic repercussions of an increase of total factor productivity in primary agricultural production. It is obvious that the opposite experiment would simply reverse the signs of the indicators and could thus indicate the likely contributions of deteriorating TFP to the overall output decline of Russian agriculture again.

Here, and in some of the following experiments, changes in TFP are simulated by increasing (decreasing) the value of the shift parameter in the production function (see equation 3-2 in chapter 3). The initial value of this indicator has been calibrated from the data set in the base period and its absolute level differs across sectors (see chapter 4). Any exogenously determined alteration in the shift parameter will therefore have a direct effect on the output in the respective sector. As we will determine the changes in TFP *ad hoc* and *exogenously* we are not analyzing the causes of such shifts. They could be due, for instance, to institutional changes which result in a more efficient use of the available stock of production factors in each sector. It also means that we are not concerned about the costs which might be associated with, for instance, a government policy that attempts to improve TFP in a given sector.

Experiment 1. The so-called widening of the 'input-output-scissors', i.e. more rapidly increasing input prices compared to output prices, in the early transition period has been made responsible for the significant output decline in Russia's agriculture (e.g. SEROVA 2000). To trace the potential effects of such changes of important input prices on the sectoral composition, Experiment 1 simulates a price increase of 10% for three intermediate products: electric power, the fuel industry, and the chemicals industry. This external shock was simulated by increasing the world market price for these commodities. Hence, this experiment is different from analyzing the effects of changes in the effective rate of protection, because not import tariffs have been used to raise domestic prices but world market price changes are paved onto domestic markets. Therefore, this experiment simulates an adverse terms-of-trade shock as it has been discussed in chapter 2. (see Figure 2-7). As mentioned above, the model is specified in such a way that capital is immobile between sectors and the wage rates are fixed. Hence, changes in relative prices should be revealed directly in the results without being compensated for or even reversed by second-round general equilibrium effects. Therefore, this model specification resembles the reality of the Russian economy more than a fully flexible CGE model in which all prices adjust to external shocks.

Experiment 2. The brief discussion of the empirical evidence on the changing role of agriculture and the food industries in chapter 1 indicated some distinct changes (see Table 1-1 and 1-2). On the eve of transition, agriculture was very capital-intensive in Russia. With the beginning of the transition period, much of the capital-intensive technology used in the large-scale collective farms deteriorated and became obsolete. It was not possible to shift the available capital stock from large-scale farms to the evolving small-scale production. Intuitively, this reduction of the collective farm sector's capital stock is one determinant of the output decline in agriculture. The quantification of the agricultural output response to such an exogenous shock, and its effects on other sectors and the economy as a whole, can be estimated with the CGE model. Hence, Experiment 2 looks into the consequences of a reduction of the capital stock in agriculture by 10%.

Experiment 3. The third experiment assumes a more positive and forward-looking perspective and looks into the general equilibrium repercussions of an increase in the TFP in agriculture by 10%. This basically corresponds to an outward shift in the supply curve and can be induced by various factors such as new production technologies, or a different institutional framework in which producers operate (cf. HENRICHSMEYER and WITZKE 1991: 256). In the Russian case, such shifts in the supply curve could be expected, for instance if the various market distortions in the food supply chain including upstream and downstream linkages were reduced (cf. MELYUKHINA and KHRAMOVA 2000). It is obvious that such changes will come about only gradually and not at zero costs. However, as we operate in a comparative static framework, we do not explicitly model the evolutionary path of such shifts but instead introduce them exogenously. Hence, by looking at an outward supply shift, this third experiment is intended to highlight the overall importance of TFP improvements in agriculture for Russia.

Simulations with 1990 model version[3]

The combined macroeconomic effects of the first three experiments (Set 1) are shown in Table 5-2. The impacts on sectoral output, producer and value added prices are shown in Table 5-3. In this table, the first column under each experiment, reports the relative output changes (ΔX); column 2 shows the changes in domestic output prices (ΔPX); and column 3 presents the changes in the price of value added (ΔPV). As discussed earlier, the 'value added price' is basically the net price which is received by producers for the sectoral good (see Table A3-1). It is calculated by subtracting production taxes and the per unit costs for intermediates from the price the producer receives. If, for instance, the prices for intermediates of one sector increase c.p., the sectoral value added price would decline, which would send out negative price signals to the producer.

Results of Experiment 1. The rising prices for electricity, fuels, and chemicals increase production costs and, with inflexible wages, lead to a reduction of employment by 0.23% and, therefore, real GDP of 0.16%. Since we assume that the Russian government can hold both real government demand and real investment constant, the total adjustment costs have to be carried by private consumption. There is also an exchange rate effect. As higher import prices lead to an increase in the absolute price of tradables, the price of home goods must fall if the overall price index is to remain at its predetermined level. Therefore, in Experiment 1, the ruble depreciates, however, moderately by 0.31%. Generally, the depreciation of the ruble should induce a restructuring of domestic production towards exports. However, this substitution effect is overcompensated for by the overall reduction of production. As a result, exports decrease slightly by 0.04%.

---

[3]   The underlying data and results of the following three simulations are discussed in more detail in WEHRHEIM (2000: 171 ff.).

Table 5-2　Macroeconomic and aggregate effects of three simulations with the 1990 version of model (Set 1): increase in import prices of important agricultural inputs (Exp. 1), decrease in capital stock in agriculture (Exp. 2) and increase in agricultural productivity (Exp. 3)

	Experiment 1[a)]	Experiment 2[b)]	Experiment 3[c)]
Experiment design:	Increase in import prices for intermediates	Deterioration of capital stock in large-scale agriculture	Increase in productivity of large-scale agriculture
Macroeconomic results:		percentage change	
Real GDP	−0.16	−1.32	1.75
Employment	−0.23	−1.16	1.55
Exchange rate	0.31	−0.57	0.78
Exports	−0.04	−0.24	0.37
Imports	−0.76	−0.29	0.43
Sectoral output:			
Industry	−0.07	−0.59	0.78
Construction	−0.10	0.21	−0.30
Food industries	−0.18	−3.95	5.22
Agriculture	−0.14	−7.05	9.74
Services	−0.14	−0.80	1.06

Notes: a) 10% increase in import prices for electric power, fuels, and chemicals. b) 10% reduction in capital stock of agriculture. c) 10% increase in agricultural productivity. d) A positive (negative) change in the exchange rate indicates a depreciation (appreciation) of the ruble.

Sectoral output reveals very different effects of higher import prices for the three sectors. As mentioned above, because of the model specification the major determinants for reallocation of resources are demand factors. On the one hand, household demand has some impact; on the other hand, the demand for intermediates is decisive for sectoral results. In the three industrial sectors, we would expect a priori rising outputs as the more expensive imported goods from the respective sectors are substituted for domestically produced inputs. Hence, demand for the domestically produced goods increases. In contrast, supply in this relatively fixed model economy can not expand easily because the sectoral capital stocks are fixed, as are the wage rates. While we observe an increase in output prices in all three sectors, output increases only in electric power and the fuel industry, decreasing slightly in the chemicals industry. This output decrease can be explained by the impact of the exogenous shock on value added prices and the respective rental rates. The chemicals industry receives approximately 60% of its intermediates from the three sectors whose import prices have risen (Table A4-4). Therefore, the consecutive increases in output prices have been more than compensated for by declines in the sectoral rental rate. In other non-food industries, output prices increase, but due to the same effect, i.e. a decrease in value added price, output in all these sectors shrinks, too. In agriculture and the food industries, output prices decline due to the reduced income in the economy, leading to lower demand for food products. In conjunction with reduced value added prices, output declines.

In the food industries the output decline is lowest in the sugar refinery sector. The share of intermediates in the total output value of the sugar industry is 68.73%. In contrast, the output decline in the flour industry (−0.20%) and the dairy industry (−0.23%) is a bit stronger as if compared with the sugar industry. Both sectors have − in the 1990 IOT − a much higher share in intermediate consumption with 91.49% (flour industry) and 93.83% (dairy industry). Nevertheless, the decline in the value added price of the sugar industry is with 0.52% bigger than the one of other food industries, for instance, of the flour milling industry. Therefore, this moderate output decline in the sugar industry can be attributed to the fact that only 0.9% of total household expenditures were spent for the good from this sector in the IOT for 1990. Hence, the reduction of income which results from the rise in unemployment hurts those sectors with high consumption shares more than those with low consumption shares.

In the case of light manufacturing, the slight increase in domestic output prices (0.10%) does not coincide with an increase in domestic output. Instead, the sector contracts by 0.24% because the value added price shrinks by 0.34% as a result of the substantial share of intermediate consumption (65.16%) in total output. Hence, those sectors with the strongest backward linkages are penalized the most.

However, the increase in prices in the three industrial sectors does not affect agricultural output significantly. Instead, an increase in prices in the three industrial sectors, which together have a share of approximately 10% of total intermediate consumption of agriculture, induces only a moderate output decline (−0.14%) in this sector.

Results of Experiment 2. A deterioration of the capital stock available in agriculture directly reduces the productive capacity of the sector compared to the reference situation and, therefore, leads to lower supply (compared to demand) and an increasing price for agricultural commodities. This will induce domestic users to change the composition of the consumption good in favor of imports, thereby reducing demand on the domestically produced good. The resulting change in the domestic price depends on the cross-price elasticity of demand for the domestic good, which itself depends on the price elasticity of demand for the composite good and the elasticity of substitution between the domestically produced good and the imported substitute. Moreover, the extent of change in the domestic price also depends on production characteristics as indicated by the elasticity of domestic supply. Low supply elasticities imply large changes of the domestic price in the case of small shifts of the demand curve. These are the most important direct effects determining the initial output price response in agriculture.

In order to estimate the final resource shifts and employment effects in the Russian economy, we must also take into account the economy-wide effects. It can be seen from Table 5-3 that a 10% reduction of the capital stock in agriculture finally yields a 7% reduction in the sector's production. With limited substitution possibilities between domestically produced and imported agricultural commodities, this will increase the price for agricultural commodities by nearly 5%. The price increase in agriculture affects input costs in other sectors. All other things being equal, those sectors with strong backward linkages to agriculture are penalized the most.

Table 5-3   Sectoral effects of three experiments (Set 1) on output, output and value added prices, simulations with 1990 version of the model

	Experiment 1[a]			Experiment 2[b]			Experiment 3[c]		
Sector	X[d]	PX	PV	X	PX	PV	X	PX	PV
1 EP	0.28	0.99	1.17	−0.64	−1.83	−2.63	0.86	2.49	3.59
2 FI	0.04	0.64	0.22	−0.19	−1.05	−0.96	0.26	1.44	1.33
3 MI	−0.04	0.28	−0.10	−0.18	−0.78	−0.52	0.25	1.06	0.71
4 CI	−0.02	0.93	−0.06	−0.52	−1.35	−1.60	0.71	1.86	2.21
5 MB	−0.07	0.19	−0.06	−0.51	−0.63	−0.46	0.68	0.85	0.62
6 WI	−0.11	0.16	−0.12	0.08	−0.33	0.09	−0.10	0.45	−0.11
7 LM	−0.24	0.10	−0.34	−1.42	−0.74	−2.02	1.82	1.03	2.61
8 CO	−0.10	0.10	−0.05	0.21	−0.28	0.10	−0.30	0.37	−0.14
9 SR	−0.08	−0.11	−0.52	−1.91	−0.65	−11.87	2.23	1.11	15.60
10 FM	−0.20	−0.01	−0.25	−5.13	1.83	−6.39	6.84	−2.08	8.65
11 MP	−0.30	−0.02	−0.93	−5.87	3.41	−17.00	7.93	−4.03	26.50
12 DP	−0.23	−0.09	−0.73	−4.09	1.69	−12.46	5.18	−1.75	17.44
13 OF	−0.12	−0.04	−0.33	−1.65	−1.09	−4.40	2.08	1.53	5.72
14 AF	−0.09	−0.02	−0.27	−6.91	−0.20	−18.70	9.40	1.18	29.64
15 AG	−0.14	−0.06	−0.33	−7.05	4.95	7.93	9.74	−5.92	−9.61
16 TT	−0.16	−0.04	−0.22	−1.00	−1.27	−1.42	1.32	1.68	1.88
17 SE	−0.13	0.08	−0.07	−0.72	−0.43	−0.38	0.95	0.58	0.49

Notes: a) 10% increase in import prices for electric power, fuels, and chemicals. b) 10% reduction in capital stock of agriculture. c) 10% increase in agricultural productivity. d) X is domestic output quantity in each sector; PX is the price of domestic output; PV is the price of value added. Names of Sectors: 1 Electric power; 2 Fuel industry; 3 Metal industry; 4 Chemicals industry; 5 Machine building; 6 Wood industry; 7 Light manufacturing; 8 Construction; 9 Sugar refinery; 10 Flour milling; 11 Meat processing; 12 Dairy processing; 13 Other food; 14 Animal feed; 15 Agriculture; 16 Trade & Transport; 17 Services.

With the exception of other food, between 50 and 70% of inputs in the food industries stem from agriculture. Obviously, price changes in agriculture will therefore affect the food industries' input costs. These effects are reflected in the discrepancies between the changes in output and value added prices of the food industries (Table 5-3). These sectors are all heavily penalized by drastically increasing prices for raw materials supplied by domestic agriculture. Even some food industries, such as flour milling, meat processing and dairy processing, which realize positive output price changes are effectively discriminated against by increasing input prices, which yield significant drops in the value added price and therefore in the sectoral rental rates. By contrast, agriculture itself benefits from lower input prices because intermediate input use in this sector is dominated by tradable inputs, the prices of which have fallen relative to domestic prices as a result of the appreciation of the ruble. Therefore, the price of value added increases by approximately 3 percentage points more than the output price.

The appreciation of the ruble can be traced back to the shift in demand. As less agricultural production leads to an increase in the absolute price of home goods, the price of tradables must fall if the overall price index is to remain at its previous value. In the capital stock simulation, the ruble appreciates slightly by 0.6%. There is also a strong employment effect that will have a differentiated impact across sectors, depending on the difference in relative factor intensities. Due to lesser investment in agriculture, overall employment decreases by 1.1%, thereby reducing real GDP by 1.3% (Table 5-2).

Results of Experiment 3. An improvement in TFP increases the productive capacity of the economy. Therefore, the results of this simulation are the mirror-image of those for Experiment 2. At a given sectoral capital stock and with given labor inputs, total factor productivity improvements will increase agricultural production and supply. The supply overhang in agriculture will cause downward pressure on the domestic price (PX) of agricultural commodities. The resulting change in domestic prices once again depends on the cross-price elasticity of demand for the domestic good, which itself depends on the price elasticity of demand for the composite good and the elasticity of substitution in use between the domestically produced and foreign good.

It can be seen from Table 5-3 that the increase in agricultural supply (ΔX) of 9.7% leads to a reduction in the price for the composite good sold by agriculture (domestically sold and exported) of 5.9%. Note also that because of substantial intra-sectoral purchases (23.6% of total costs of production) the net price (PV) decreases even more (9.6%). On top of these direct effects on agriculture, there are indirect effects which determine the final resource allocation in the Russian economy. With the exception of the wood industry and construction, all sectors benefit from the increasing demand that results from increased labor demand (1.5%), the increase of real GDP (1.8%; see Table 5-2) which effectively increases the available income in the economy.

Furthermore, in this relatively inflexible model economy, the results of this experiment are quite illustrative for Cochrane's treadmill theorem: innovative farmers increase productivity and, in doing so, increase output. Because of relatively low CET values only a small share (approximately 1%) of total agricultural output can be disposed of on the world market (exports of agriculture increase by 25%). Domestic demand does not increase substantially and, therefore, agricultural prices are forced to decline. Even though these effects are obvious, they have to be understood by farm lobbyists. The real exchange rate depreciates again and total exports rise. The increase in income available in the economy overcompensates for the depreciation of the exchange rate and causes imports to increase. Higher demand for intermediates in the food industry sectors results also in higher demand for industrial intermediates which c.p. causes an increase in the output prices in these sectors. As the value added price in these sectors does not decline, they become more competitive.

Simulations with 1994 model version

The three experiments discussed in the last section were simulated in a similar way with the 1994 version of the model. However, the results are likely to differ not only because of the different data base but also because of different trade elasticities. In particular, the constant elasticity of transformation on the production side was set in the updated model lower than that chosen in the previous model. This was done in order to specify the model more clearly as a short-run model. With respect to the data base, one of the most important changes between the two model versions is the fact that agricultural production in the updated model is split into small-scale and large-scale production. This has some implications for the experiment design: while all three experiments were simulated in a similar way to those using the 1990 model, some variations seemed useful. In Experiment 1, one variation in the experiment design was to additionally increase the import prices of the mechanical engineering or machinery sector by 10%, because this sector also delivers important intermediates for agricultural production. In Experiment 2, the reduction of the capital stock (by 10%) and in Experiment 3 the increase of productivity (by 10%) were only applied to the two large-scale agricultural sectors. The argumentation behind this difference in the experiment design is the fact that it was mainly large-scale agriculture which suffered from the deterioration of the capital stock. In addition, technology improvements induced by programs implemented by international donors or the Russian government are likely to focus on large-scale farms. In such cases, an outward shift of the supply curve (as simulated in Experiment 3) would occur in the large-scale farm sector only. As long as small-scale agriculture in the Russian Federation continues to be neglected by policy-makers as well as by other economic agents in agri-business – including, for instance, financial intermediaries who could provide the required credit for technology improvements – it is to be expected that the productivity in the small-scale sector will not be enhanced in the mid term.

The aggregate results of the three experiments are presented in Table 5-4 and the sectoral results in Table 5-5, each in the same format as the results obtained with the 1990 version of the model (see Tables 5-2 and 5-3). Generally, the results point in a similar direction, even though the effects are mostly more moderate. However, some distinct differences deserve comment.

Results of Experiment 1. Generally, the increase in world market prices for the goods of four industrial sectors simulated with the 1994 version of the Russia model also has a similar impact on the Russian economy compared with the results of the same experiment simulated with the 1990 version of the model. Additionally, the results of this moderate adverse terms-of-trade shock are along the same lines as what has been hypothesized with the graphical presentation of the Keynesian model economy in Figure 2-7 (see chapter 2): The economy contracts slightly – GDP declines by 0.4% – and unemployment occurs. Hence, the economy adapts to the external shock by operating below its production possibility frontier.

Differences in the external responses to this terms-of-trade shock are due to a stronger depreciation of the ruble in this experiment (0.86% instead of 0.31%), which yields a stronger reduction in exports but even more so of imports. Due to the increase in import prices, imports are reduced quite significantly (–3.71%). However, not all imports are replaced, which means that the imported intermediates become more expensive because of which the economy contracts and reduces employment (by 0.41%).

In sectors in which the price for the imported good has been increased exogenously, domestic output should increase as the domestic good has become less expensive relative to the imported good (see Table 5-5). In fact, imports are reduced in the four sectors which were exposed to an increase of import prices between 4.5 and 8.5%, while imports in other non-food industries are reduced only up to 2% (light manufacturing).

In a similar way to the respective experiments that were carried out with the 1990 version of the model, the capacity to restructure in this inflexible model economy is limited, which means that the chemicals industry does not experience an increase in output even though the output price (ΔPX: 1.6%) rises (see Table 5-5). The main explanation for the output contraction (ΔX: –1.1%) is the fact that the rise in import prices increases the costs for the intermediates used in the production process of this sector, which means that the value added price declines (ΔPV: –2.1%; not reported in Table 5-5).

The same mechanism, the increase of import prices, also causes the food industries and the two large-scale agricultural sectors to deteriorate. The value added price is reduced in all food industries between 0.7 and 1.2%, which contributes to the reduction of output in these sectors. At the same time, the depreciation of the exchange rate causes the import-dependent food industries to reduce imports significantly. The sum of these sectoral reductions in imports yields an economy-wide reduction of imports of 3.7%. However, the restructuring towards exports is not strong enough to compensate for the contraction in import-dependent sectors, which means that overall domestic output contracts.

The large-scale crop sector reduces production only slightly (ΔX: –0.35%), while the livestock sector experiences a more significant decline in output (ΔX: –1.47%). In both sectors, the value added price is reduced because of the increasing price for intermediates which results directly from the increase in import prices in the four industrial sectors. However, this price increase is neither strong enough nor was the share of intermediates from these sectors (as revealed by the IOT for 1994 in Table A4-5) in both large-scale agricultural sectors large enough to do much harm to either sector.

Table 5-4   Macroeconomic and aggregate effects of three simulations (Set 2) with the 1994 version of the model: increase in import prices of important agricultural inputs (Exp. 1), decrease in capital stock in large scale agriculture (Exp. 2) and increase in productivity in large scale agriculture (Exp. 3)

Experiment design: Macroeconomic results:[d]	Experiment 1[a] Increase in import prices for intermediates	Experiment 2[b] Deterioration of capital stock in large-scale agriculture percentage change	Experiment 3[c] Increase in productivity of large-scale agriculture
Real GDP	−0.26	−0.21	0.38
Employment	−0.41	−0.08	0.11
Exchange rate[d]	0.86	−0.25	0.56
Exports	−0.26	−0.13	0.30
Imports	−3.71	−0.15	0.34
Sectoral output:			
Raw material sector[e]	−0.16	−0.09	0.21
Other industries	0.99	−0.19	0.41
Construction	−1.10	0.01	−0.02
Food industries	−0.42	−1.42	2.67
Agriculture, of which[e]			
Large-scale agriculture	−0.61	−2.83	6.05
Small-scale agriculture	−0.48	−0.63	1.35
Services	−0.20	−0.01	−0.09

Notes: a) 10% increase in import prices for electric power, fuels, chemicals and machinery sector. b) 10% deterioration in capital stock in LAE (crop and livestock sectors). c) 10% increase of TFP in LAE (crop and livestock sectors). d) A positive (negative) change in the exchange rate indicates a devaluation (appreciation) of the ruble. e) Here and in the following Tables we report the aggregate effects for the export-oriented 'raw material sectors', comprising sectors 1–4 (electricity, fuel, metal, chemicals) and sector 6 (wood). The aggregate 'other industries' consists of sectors 5 (mechanical engineering) and 7 (light manufacturing). The aggregate 'large-scale agriculture enterprises' (LAE) refers to the large-scale farms producing crop and livestock products as well the animal feed sector. 'Small-scale agriculture' consists of the private subsidiary plot (LPH) sector and the private farm sector. The aggregate 'services' embraces the trade and transport sector as well as other services.

Table 5-5   Sectoral effects of three experiments (Set 2) on output, output and value added prices, and imports, simulations with the 1994 version of the model

	Experiment 1[a] Increase in import prices			Experiment 2[b] Deterioration of capital stock			Experiment 3[c] Increase in productivity		
Sector	X^d	PX	M	X	PX	PV	X	PX	PV
1 EP	−0.21	0.31	−6.23	−0.15	−0.51	−0.76	0.31	1.07	1.61
2 FI	0.07	0.97	−4.80	−0.04	−0.26	−0.26	0.10	0.58	0.59
3 MI	−0.05	0.61	−0.55	−0.06	−0.20	−0.19	0.16	0.43	0.51
4 CI	−1.12	1.62	−4.56	−0.13	−0.30	−0.25	0.28	0.63	0.54
5 MB	1.78	2.80	−8.52	0.01	−0.09	0.01	−0.03	0.16	−0.04
6 WI	0.08	0.17	−0.71	−0.18	−0.25	−0.23	0.39	0.53	0.50
7 LM	−0.32	0.03	−2.08	−0.52	−0.39	−0.86	1.12	0.90	1.88
8 CO	−1.10	−0.37	-	0.01	−0.05	0.01	−0.02	0.09	−0.01
9 SR	−0.31	−0.22	−1.97	−1.53	0.42	−5.91	2.51	0.37	10.31
10 FM	−0.82	0.02	−2.07	−3.76	2.68	−5.95	6.73	−4.13	10.98
11 MP	−0.37	−0.39	−2.24	−0.63	−0.14	−2.89	1.50	0.13	7.14
12 DP	−0.25	−0.43	−2.19	−0.47	−0.31	1.45	1.13	0.55	3.52
13 OF	−0.24	−0.21	−2.22	−0.62	−0.19	−1.93	1.11	0.51	3.51
14 AF	0.64	0.55	-	2.26	3.02	5.89	−7.24	−7.40	−17.50
15 CP	−0.35	0.11	−1.49	−2.86	5.16	8.67	4.92	−8.15	−13.65
16 LP	−1.47	−0.65	−3.05	−4.84	1.59	2.68	12.96	−4.39	−7.94
17 LPH	−0.59	0.04	-	−0.62	−0.64	−0.94	1.35	1.40	2.07
18 PF	−0.42	−0.19	-	−0.74	−0.66	−1.16	1.44	1.24	2−29
19 TT	−0.19	−0.26	−0.79	−0.15	−0.26	−0.28	0.26	0.47	0.49
20 SE	−0.21	−0.04	-	0.09	0.05	0.06	−0.34	−0.17	−0.22

Notes: a) 10% increase in import prices for chemicals, metal, fuel industry and in machinery; b) 10% reduction in capital investments in agriculture. c) 10% increase in total factor productivity in large-scale agriculture. d) X = domestic output quantity in each sector; PX = price of domestic output; PV = price of value added. M = Imports. Names of Sectors: 1 Electric power; 2 Fuel industry; 3 Metal industry; 4 Chemicals industry; 5 Machine building; 6 Wood industry; 7 Light manufacturing; 8 Construction; 9 Sugar refinery; 10 Flour milling; 11 Meat processing; 12 Dairy processing; 13 Other food; 14 Animal feed; 15 Large-scale farms: crop production; 16 Large-scale farms: livestock production; 17: Small-scale farms: private subsidiary plots (LPH); 18 Small-scale farms: private farmers; 19 Trade & Transport; 20 Services.

In a similar way to all other sectors, the depreciation also induces a reduction of imports in both large-scale agricultural sectors. As the overall import share in the base period was lower in the crop sector compared to the livestock sector (8.3 and 12.8%, respectively), the absolute reduction of imports in the large-scale livestock sector is much greater. As the expenditure shares of the average Russian household in 1994 were low for both large scale agricultural sectors (0.2 and 2.5%, respectively; see Table 4-6), the reduction of domestic absorption in this experiment did not stimulate the output decline of large-scale agriculture directly.

The increase in import prices, the depreciation of the exchange rate and the concomitant contraction of the economy meant that households have less income to spend. In contrast to large scale agriculture, the agricultural sectors comprising the private subsidiary plots and the private farmers contract almost at the same rate, even though they buy much fewer intermediates from the sectors in which import prices rose. At 10.5% of total expenditures spent on the good from the private subsidiary plot sector, this share is the highest for all primary agricultural sectors (see Table 4-4). Therefore, in the case of the two small-scale agricultural sectors, it is not so much the effect of increasing import prices for intermediates but rather the high expenditure share the average Russian household in the model spends on buying the good from these two sectors which explains the sectoral output decline. Hence, while the relatively low intensity in using industrial intermediates insulates the small-scale farm sectors, and here particularly the private subsidiary plots, its important role with respect to household expenditures keeps these sectors exposed to economy-wide developments.

Results of Experiment 2.   The results of Experiment 2 simulated using the 1994 model version comply by and large with those of the 10% reduction of the capital stock using the 1990 version of the model. While the sign of changes of all indicators is the same, the absolute size of the respective changes is mostly more moderate. This is due to the fact that only the two large-scale agricultural sectors were exposed to the exogenous shock, which reveals the actual situation in the transition period much more realistically. The reduction in the productive capacity in large-scale agriculture yields a small contraction in domestic output (ΔGDP: -0.21%). Holding the domestic price index constant, the exchange rate appreciates only slightly (by 0.21%), which again results in a moderate decline of imports and exports.

Restructuring of primary agriculture.   The sectoral results of this experiment can be interpreted as a *validation of the model* in that they *replicate one real world phenomenon*: while the output from the large-scale farms declined in the transition period significantly, the role of small-scale agricultural production increased, at least in relative terms. Obviously, various factors were driving this development, but we offer here an additional explanation that is often overlooked: while the capital stock of the former *kolhkozi* and *sovkhozi* continued to deteriorate, this factor did not in fact affect the small-scale sector so negatively. The results indeed show the effects that would be expected and which have been observed in reality: a strong decline in output in the large-scale sector (ΔX: -2.83%) coincides with a less pronounced decline in the small-scale sector (ΔX: -0.63%). Speaking in relative terms, the share of small-scale agriculture grew, which effectively implied significant restructuring of primary agricultural production that has been observed in Russia and in many other CIS countries. In fact, small-scale agriculture has partially compensated for the output decline in large-scale agriculture by maintaining production more or less constant.

Therefore, the role of these small-scale producers should neither be underestimated nor should one conclude that Russia's agricultural production must rely on the small-scale farms only. Instead, it is important to *identify the most efficient small-scale producers and help them to commercialize their operation by entering more formal food markets.* A study for the Ukraine showed that, even though the majority of LPH production is a subsistence activity, there is a small group of these producers who exploit their resources to the full potential. These farm households have the potential to become commercial small farmers (PERROTTA 1999: 29).

Effects on the food industry.   As a result of the deteriorating capital stock and the consecutive decline in large-scale crop and livestock production, the output price for the composite good produced by both sectors increases (*ΔPX: 5.16 and 1.59%, respectively*). In turn, this contributes to a rising value added price in four out of the five food industries and to the respective output decline in these sectors. The share of intermediates bought from large-scale agriculture, and more specifically from the large-scale crop sector is highest in the flour milling sector (more than 60% of total production costs), followed by the sugar refinery sector, which spends more than 45% of its total expenditures on primary goods stemming from the large-scale crop sector (see Table A4-5). Therefore, the increase in the output price in the large-scale crop sector reduces the value added price in these two food industries most significantly, which leads to the most negative output responses compared to other food industries.

The output prices for the industrial sectors also decline, which in this demand-driven Keynesian model economy is mainly the result of reduced demand for intermediates from the agro-food sectors. The respective decline in output is strongest in the two sectors in which the decline in the output prices is associated with an even greater decline in the value added price, namely in the energy and light manufacturing sector.

Effects on the exchange rate.   The results of the phenomenon explain and validate yet another phenomenon characteristic of the early transition period: in the period between 1992–94, only the nominal exchange rate of the Russian ruble against western currencies was devalued, while the real exchange rate appreciated. In 1993, the real appreciation of the ruble against the US dollar was about 110% and about 38% in 1994 (POGANIETZ 2000: 142). In fact, this simulation offers another plausible explanation for this trend: the declining capital stock and the concomitant decline in output translated into higher prices for home goods. Under a system with a flexible exchange rate, this resulted in an appreciation of the ruble against other major currencies, making imported food items relatively more competitive in that period. In fact, in this simulation, the flour milling sector and the agricultural crop sector both react to the decline in domestic supply by increasing imports (*ΔM: 0.56% and 5.37%, respectively*).

Against this background, it is important to *acknowledge two important links between the agro-food sectors and macroeconomic developments* which were shown with our model economy: first, the declining competitiveness of domestic agriculture contributed to the appreciation of the real exchange rate; second, this appreciation made imported food commodities, relatively speaking, more competitive. It should therefore be understood that not only the liberalization of trade regimes but also domestic economic reasons were responsible for the increased competitiveness of imported food commodities and the respective sizable inflow of imported food in the early transition period.

Results of Experiment 3. The third experiment again assumes a more positive perspective by exogenously introducing an outward shift in the supply curve. Reversing the sign of the simulated changes would provide an indication of the opposite experiment, namely an inward shift in the supply curve. The 10% increase in TFP is simulated in the two large-scale agricultural sectors only. In the data base these two sectors produced roughly 50% of total agricultural output only. Hence, in contrast to the same experiment carried out using the 1990 version of the model, only half of agricultural production benefits from this productivity increase.

Macroeconomic effects. Obviously the outward shift in the supply curve in the large-scale agricultural sectors only limits the positive effects of this optimistic scenario as if compared with the effects of the same experiment using the 1990 version of the model (Experiment 3 of Set 1): first, GDP increases only by 0.38% in this experiment, as compared to 1.75% in the same experiment carried out with the older data base. Second, the employment effect is also much less positive compared to the same experiment simulated with the older model version. While most sectors hire additional labor due to the output expansion, both large scale agricultural sectors together release 4% of the original labor force due to the positive development of TFP in these sectors. Third, this experiment yields a more moderate depreciation of the ruble as compared to the same experiment carried out with the 1990 version of the model.

Sectoral effects. Output growth is obviously substantial in the two large-scale sectors (Table 5-5). It is noteworthy that this increase is far more distinct in the large-scale farm sector specialized in livestock production than compared to the crop sector (ΔX: *12.96 and 4.92%, respectively*). This is due to the fact that the experiment results in a more moderate decrease of the value added price in the livestock sector compared to the crop sector (ΔPX:*−7.94% and −13.65%, respectively*). A somewhat counterintuitive result is obtained for the animal feed sector, which experiences a significant output decline despite the fact that the large-scale livestock sector increases output. This seems to be due to the fact that this sector is a non-tradable sector and also only contributes about 0.1% to GDP at factor costs. Due to the flexible labor market, the reduction in the value added price by 17.5% yields an even greater release of labor (by −23.48%), which in the end contributes to the reduction in output in this sector by more than 7%. Hence, the response of this sector should be interpreted cautiously.

Degree of commercialization. As a consequence of the close relationship we have modeled between both small-scale farms sectors (LPH and private farms) and the two large-scale farm sectors via input-output relationships, these two sectors also benefit from the growth in the two large-scale farm sectors and expand production (*ΔX: 1.35 and 1.44%*). As both sectors produce non-tradables, the slightly larger increase in the private farm sector must be caused by a more significant increase in domestic demand for the good produced by private farms. As the household expenditure share for the good from the LPH sector is higher than that from the private farm sector, the higher degree of commercialization of the private farm sector must be responsible for the marginally more positive supply response in this sector. In fact, it is the growth in the food industries which induces additional intermediate demand for the good delivered by the private farm sector. Hence, this sector benefits slightly more than the LPH sector, which has fewer commercial links with the processing industries. Even though the difference is marginal and the underlying empirical assumption that in 1994 the private farms sector in fact had better commercial links than the LPH producers can be disputed, the relevant point is the underlying causality: the *higher the degree of commercialization* of a specific segment of agricultural production, *the more this sector will benefit from economy-wide growth.*

Effects on food industries. Output expands in all food industries because the price for the good from the domestic large-scale agricultural sectors decreases because of which the most important intermediates of these sectors become less expensive (see Table 5-5). In the flour milling and sugar refinery industries, the effects are particularly positive, and output growth is clearly favored by the strong decrease in the output price in the large-scale crop sector, increasing the value added price in these two food industries by more than 10%. Given the fixed nominal wage rates, and as labor demand is determined endogenously, both sectors hire substantially more labor, contributing to the overall positive employment effect of this experiment (*ΔL: 13.01 and 18.44%*).

Effects on non-food sectors. In the non-agricultural sectors the effects of this experiment are positive, with the only notable exception of the non-tradable service sector. However, the positive effects on the industrial sectors are generally much more moderate. In the mechanical engineering and construction sector, output remains more or less constant. The highest increase in output is experienced by the light manufacturing sector, which grows by 1.12%. Again, in our Keynesian model economy, it is final demand which is responsible for this positive effect. On the one hand, it is the increased demand for intermediates from light manufacturing, but even more so the higher demand by private consumption which is due to the increase in income available in the economy. Light manufacturing, at 14.1%, is the non-food sector with the highest share of household expenditures and it therefore benefits most from the income increase that was most significant in the agro-food sectors (see Table 4-4). In fact, by reversing the experiment an additional validation of the model can be shown: a decrease in agricultural productivity and, hence, a concomitant strong output decline would, speaking in relative terms, induce the highest production decline in non-agricultural sectors in light manufacturing. This relatively strong decline of light manufacturing in the first years of Russia's transition period has been documented in Figure 4-2.

Summing up, the results obtained from the 1994 experiments (Set 2) produce similar effects for the various sectors in the model economy compared with the results obtained from the 1990 model (Set 1). The higher level of disaggregation allows for more specific conclusions, particularly with respect to the dichotomy between small-scale and large-scale agriculture. The results obtained from experiments conducted with both model versions highlight the two major contributions agriculture has to make to general economic development: first, the increase in supply results in decreasing consumer prices for food and, given the high expenditure share Russian households spend on food, contributes significantly to an increase in consumer welfare; second, as factor productivity increases labor is released from large-scale agriculture and could be used in other sectors once full employment is reached.

Experiment design for effects of sector-neutral factors

The last two sections presented three agricultural sector simulations, each carried out with the older and the newer model version. In the following section, we will present results of simulations which are complementary to the experiments conducted with the 1994 version of the model. The focus of these experiments will be more on sector-neutral factors and on their contribution to the decline in output in Russia's agro-food sectors.

With the first experiment, we will look into the effects of an alteration in the total factor productivity in the transport sector. The second experiment will ask what happens if such negative effects sector-neutral effects are combined with negative effects in the sector itself and, hence, sector-specific effects. The third experiment will be an extension of the terms-of-trade experiment in the previous section. This time, however, all tradable sectors of the Russian economy will be exposed to a negative terms-of-trade shock.

Transport sector.   The discussion in chapter 2 indicated that the performance of the agro-food sector in the transition period was driven by various factors. One argument was that because of the vastness of the country and the long distances between production and demand locations the condition of the transportation sector is of particular relevance for Russia's agro-food sector. Therefore, in the following we will discuss the effects of varying performance levels of Russia's transport sector and its effects on the agro-food sectors. The overall importance of the sector in Russia's economy is already indicated by the figures given in Table 4-3. In 1994, the contribution of this sector to gross production in Russia's economy was about 13% while the sector's share in GDP was with about 18% even higher. Almost 70% of 'the good' sold by the sector was used as an intermediate input in the Russian economy, about 20% of the sectoral good was demanded by the government, and only 10% of it was sold to households (see Table 4-5 to 4-8).

The first experiment looks into the effects of a reduction in the TFP of the transport sector by 20% (see above for a discussion of this parameter). In doing so, we assume that both the weak maintenance of the sector's infrastructure and low motivation of the labor force in this sector contributed to the deterioration in TFP in this sector in the course of transition. The second experiment looks into the effects of a combination between sector-neutral and sector-specific negative shocks on the agro-food sectors. The experiment complements the 20% reduction in TFP in the transportation sector with a 50% reduction in direct domestic support granted in the base period to large-scale agriculture. As indicated in Table 4-7, this support to agriculture is modeled via negative indirect tax payments. Direct government support has been granted to large-scale farms throughout the transition period, either as direct production subsidies to livestock farms or as input subsidies for crop farms. However, the amount of such support declined continuously in the transition period due to the severe government budget deficit. Therefore, this experiment simulates what happens if both events take place simultaneously.

Inefficiencies in the transport sector

Experiment 1.   The results of Experiment 1 indicate the central role of the transport sector for the whole Russian economy. The macroeconomic effects of this shock are generally negative (see Table 5-6). The decline in the sector's TFP results in a reduction of output in the sector itself and thereby increases the price of the sector's output, which is an important input for all other sectors in the economy. Given the Keynesian model economy, the decline in TFP in one sector is associated with a sizable reduction of labor (–7.2%). Therefore, less income is available in the economy, also reducing demand. The inward shift of the production possibility frontier (see Figure 2-6 in chapter 2) is revealed by a decrease in real GDP of almost 8%. The reduction in total domestic output causes also an appreciation of the exchange rate by 7%. Therefore the reduction in the productive capacity yields also a contraction of both exports and imports.

Sectoral effects.   This experiment yields very high negative effects across all sectors. However, the changes in output of sectoral aggregates as shown in Table 5-6 already indicate that the extent of the decline is quite different across sectors, and particularly strong in the agro-food sectors. The disaggregated sector results in Table 5-7 show that the variation of output results is even greater. The transport sector is not the sector that is affected most negatively. Because of the strong decline in supply the price for the composite good produced by this sector increases which also translates in an increase of the value added price (ΔX: -11.86%; ΔPX: 42.62, and ΔPV: 49.86). This increases the production costs for all sectors which use 'the good' from the transport sector as an input, which again induces reductions in the value added price of all sectors, except in the transportation sector itself. Hence, output in almost all sectors declines. In particular, this increase in the costs for transportation most negatively affects the agro-food sectors that produce bulk commodities, for example the large-scale crop sector, flour milling, and the animal feed sector. Next to the animal feed sectors, the output reduction is most significant in the meat processing industry and the flour milling sector (ΔX: -19.41 and -15.24%). The tremendous decline in the value added price in the agro-food sectors additionally proves how severe the effects of this shock, which originally occurred outside of the agro-food sectors, are for these sectors' economic development.

Experiment 2. The transition process did not take place in a nicely planned fashion. Hence, sectoral reforms in the agro-food sector overlapped with economic developments elsewhere. The effects of sectoral agricultural policies often became more pronounced because of developments in other sectors. While the first experiment showed the significant effects of the deterioration in efficiency in the transport sector, the second experiment adds a 50% decline in direct support granted in the base period to the former collective farms.

Macroeconomic effects. The differences in results between Experiment 1 and 2 can be attributed only to the additional shock to which the economy is exposed with Experiment 2: the decline in direct domestic support granted to large scale agriculture. One of the most important observations here is that these differences between the results of the two experiments are not all that great, again highlighting the outstanding importance of the transport sector. The reduction in GDP, employment, the appreciation of the exchange rate, and the reduction in exports and imports in Experiment 2 is just slightly higher than the respective change in Experiment 1.

Sectoral effects. Notable differences between the two experiments become obvious only when the sectoral results are examined. Due to the additional reduction in domestic support to large-scale agriculture, the marginal reduction is most significant in this sector, followed by the food industries (ΔX: -14.09% and 15.35%, *respectively*). Additionally, the disaggregated results in Table 5-7 show that, in comparison to Experiment 1 of Set 3, the output decline is in all agro-food sectors more negative. The only exception is the animal feed sector, where the output price declines less compared with Experiment 1 as a result of the significant reduction in demand from the livestock sector. This sector reduces output in Experiment 2 by 11.2% only, compared to a moderate decline in Experiment 1 of 2.2%. In turn, this results in a somewhat more moderate decline in the value added price of the animal feed sector, which feeds back into a lower reduction compared with Experiment 1.

In the two large-scale agricultural sectors, the reduction of domestic support results in a lower decline of the output price or even increases the output price in the livestock sector. As these two sectors deliver high shares of their products as intermediates to the food industries, the decline of the value added price in the food industries is higher compared to Experiment 1, which explains the stronger output decline in these sectors.

In the second experiment, the effects on small-scale agriculture are also significantly lower compared to large-scale agriculture. What might be surprising, however, is the fact that the two small-scale sectors show such significant effects at all. Since both sectors spend only a small share of total production costs on the transportation sector, the negative development in these two sectors can not be driven by this sector's poor performance only. Instead, here again it is the strong decline of income generated by both experiments which particularly hurts those sectors which have a high household expenditure share and – as mentioned before – this is particularly the case for small-scale agriculture.

Table 5-6 Macroeconomic and aggregate effects of three simulations (Set 3): reduction in total factor productivity in the transportation sector (Exp. 1), reduction in total factor productivity of transportation sector and 50% reduction of direct domestic government support to large-scale agriculture (Exp. 2), and increase in import prices in all sectors by 10% (Exp. 3)

Experiment design	Experiment 1[a)] Reduction in TFP in transport sector (−20%)	Experiment 2[b)] Reduction in TFP in transport sector (−20%) and reduction in domestic support to LAE (−50%) percentage change	Experiment 3[c)] Increase in import prices in all sectors (+10%)
Macroeconomic results:			
Real GDP	−7.72	−7.95	−0.47
Employment	−7.22	−7.72	−0.59
Exchange rate[d)]	−6.98	−7.36	0.83
Exports	−7.13	−7.32	−0.61
Imports	−7.61	−7.82	−8.46
Sectoral output:			
Raw material sectors[e)]	−6.35	−6.48	−0.29
Manufacturing	−8.36	−8.64	0.71
Construction	−3.59	−3.55	−1.33
Food industries	−13.74	−15.35	−0.28
Agriculture, of which			
Large-scale agriculture	−10.82	−14.09	0.06
Small-scale agriculture	−9.62	−10.54	−0.91
Services	−8.97	−8.90	−0.26

Notes: a) 10% reduction in total factor productivity of transportation sector (Exp. 1). b) 10% reduction in total factor productivity of transportation sector and 50% reduction in direct domestic government support to large-scale agriculture (Exp. 2). c) 10% increase in total factor productivity of transportation sector and in large-scale agriculture (Exp. 3). d) A positive (negative) change in the exchange rate indicates a devaluation (appreciation) of the ruble. e) See Table 5-3 for the sector aggregate, for which results are reported in the following.

Table 5-7   Sectoral effects of three simulations (Set 3) on output, output and value added prices and imports

Sector[b]	Experiment 1 Decrease in total factor productivity in transport sector by 10%			Experiment 2 Decrease in total factor productivity in transport sector by 10% and reduction in direct support to agriculture by 50%			Experiment 3 Increase in import prices by 10 % in all sectors		
	X[a]	PX	PV	X	PX	PV	X	PX	M
1 EP	−6.81	−11.46	−30.78	−7.02	−12.11	−31.63	−0.34	−0.02	−6.61
2 FI	−2.09	−7.37	−11.92	−2.16	−7.76	−12.29	0.06	0.93	−4.93
3 MI	−9.69	−3.51	−27.53	−9.78	−3.80	−27.76	−0.39	1.35	−5.01
4 CI	−7.74	−6.75	−14.13	−7.93	−7.16	−14.45	−1.28	1.57	−4.82
5 MB	−6.58	−2.56	−9.36	−6.55	−2.68	−9.21	1.60	2.77	−8.69
6 WI	−9.61	−3.64	−12.13	−9.86	−4.00	−12.43	0.06	0.64	−5.71
7 LM	−11.32	−6.50	−18.16	−12.11	−7.11	−19.37	−0.76	−0.16	−2.87
8 CO	−3.59	4.13	−2.08	−3.55	4.04	−2.05	−1.33	−0.34	−6.46
9 SR	−8.32	−10.92	−29.07	−9.83	−10.80	−33.60	0.38	2.17	−11.16
10 FM	−15.24	−5.66	−23.23	−18.14	−3.82	−27.38	−0.85	1.92	−12.63
11 MP	−19.41	−1.36	−63.16	−21.16	−0.57	−66.71	0.05	0.96	−13.11
12 DP	−10.41	−9.76	−28.81	−11.66	−9.72	−31.84	0.32	1.14	−12.63
13 OF	−10.48	−9.34	−29.37	−11.19	−9.71	−31.11	−0.42	1.15	−13.19
14 AF	−25.91	−16.85	−53.57	−19.62	−12.36	−42.81	1.67	2.88	-
15 CP	−12.66	−5.26	−13.69	−14.56	−1.64	−15.73	0.78	0.95	−12.49
16 LP	−2.24	−0.38	−1.07	−11.21	5.26	−5.48	−1.53	−0.96	−15.05
17 LPH	−9.54	−12.54	−14.25	−10.46	−13.45	−15.60	−0.93	0.06	-
18 PF	−10.73	−11.82	−16.44	−11.61	−12.51	−17.73	−0.54	0.54	-
19 TT	−11.86	42.62	49.86	−12.05	42.06	49.25	−0.22	−0.28	−5.44
20 SE	−6.93	−2.65	−4.56	−6.68	−2.53	−4.39	−0.28	0.08	-

Notes: a) X = domestic output quantity in each sector; PX = price of domestic output; PV = price of value added. M = Imports. b) Names of Sectors: 1 Electric power; 2 Fuel industry; 3 Metal industry; 4 Chemicals industry; 5 Machine building; 6 Wood industry; 7 Light manufacturing; 8 Construction; 9 Sugar refinery; 10 Flour milling; 11 Meat processing; 12 Dairy processing; 13 Other food; 14 Animal feed; 15 Large-scale farms: crop production; 16 Large-scale farms: livestock production; 17: Small-scale farms: private subsidiary plots (LPH); 18 Small-scale farms: private farmers; 19 Trade & Transport; 20 Services.

Overall decline of Russia's terms of trade

Experiment design.   The third experiment of set 3 is similar to experiment 1 of Set 1 and Set 2, which simulated a *partial* deterioration of Russia's terms of trade. In the recent past, however, Russia has experienced negative terms-of-trade shocks which actually affected the whole economy. Such events were associated with periods during which the real exchange rate of the ruble against western currencies was devalued. In the years between 1992–94, the price differential for consumer products in Russia and in OECD countries changed by a factor of five due to the high level of inflation in Russia and an inability to stabilize the macroeconomy. In the same period, the ruble devalued against major western currencies only by a factor of two. Therefore, the real exchange rate of the ruble against these currencies had appreciated sharply by a factor of about 2.5 (POGANIETZ 2000: 141). In the period between 1994 and 1997 the appreciation of the real exchange rate continued but at a much lower rate (CENTRAL BANK OF RUSSIA 1998). Then in late 1997 the financial crisis started to harm economic development in various south-east Asian economies. In the first half of 1998 Russia's terms of trade started to deteriorated because of the financial crisis. Later on in 1998, the worsening of the country's terms of trade was complemented with a 'contagious effect': international investors perceived an increased risk to invest in emerging economies and the trust in the Russian economy deteriorated in spite of positive signs of recovery in 1997. Therefore, a distinct and sudden depreciation of the real exchange rate was experienced by Russia right after the financial crisis culminated in mid–1998. On the one hand, it was expected that the consecutive real devaluation of the ruble would increase the price competitiveness of domestic agro-food sectors and would open a 'window of opportunities'. On the other hand, it was argued that – at least in the short term – mainly export-oriented sectors would benefit from a devaluation of the ruble, while import-dependent sectors, among them foreign companies and joint ventures in the food industry, which rely on imported raw materials would contract as inputs become more expensive (SEROVA 1998). However, the exogenous shock which initiated the financial crisis was not a devaluation but a negative terms-of-trade shock. Therefore, in the following, we simulate a much more moderate adverse terms-of-trade shock by exogenously increasing the import prices of all tradables by 10%.

Macroeconomic effects.   Given the flexible exchange rate in this model, a deterioration in the terms of trade should be at least partially offset by a depreciation of the real exchange rate. Similarly to the response of the ruble to the financial crisis this is indeed the case: the exchange rate depreciates in our simulation by 0.83% (see Table 5-6). However, the depreciation following this adverse terms-of-trade shock is much more moderate than the one actually experienced in 1998. But as stated above, in reality the negative terms-of-trade shock was re-enforced by a 'contagious effect' because of which the financial crisis became so severe and the devaluation was so significant. Such psychological factors are clearly outside the scope of our model. In our model economy, the immediate response to the adverse terms-of-trade effect was that the overall demand of the Russian economy for imports in domestic currency, and hence in nominal terms, is reduced by 8.46%. The rising prices for imported inputs can – at least in this inflexible, short-run economy – not induce sufficient import substitution and therefore does not induce overall growth but instead yields a decline in GDP of –0.47%. Again, those sectors which are hurt most by the increase of import prices react by releasing labor which yields – similarly to the effects shown in Figure 2-7 – an increase in unemployment.

Sectoral effects. The increase in import prices has negative effects for various reasons: first, it affects those sectors which, in the base period, imported significant shares of their inputs; second, the devaluation reduces the available income in the economy. As a result of the first argument, the raw material sectors and the food industries suffer and reduce output. As a result of the second argument, the private subsidiary plot sector, for instance, reduces output by almost one percent. In contrast, the aggregated large-scale agricultural sector increases production slightly, while the positive output response is more pronounced in domestic manufacturing (see Table 5-6). A closer look at the specific sectoral results in Table 5-7 reveals that the positive effect in large-scale agriculture originates from the positive development in the large-scale crop sector which restructures part of its production towards exports. The higher import share of the large-scale livestock sector in the base period (12.8% of domestic production see Table 4-8) actually induces a decline of output in this sector.

The results are also more mixed within the food industries, in which only the flour milling sector and the sector 'other food' reduce output, whereas the other three sectors expand production. The increase in the meat processing and the dairy industries is induced by an increase in the sectoral value added price, which in these two sectors is caused by the declining output price in the livestock sector (ΔPX: – 0.96%).

Similarly, the results are also more mixed in the various raw material sectors. The export-oriented fuel industry increases output slightly because of the increased price competitiveness, while particularly the chemicals industry, with an import share of 47.9% in total production, experienced significant output declines. Furthermore, the last column in Table 5-7 shows the developments of sectoral imports in response to the increase in import prices. While imports decline under-proportionally in the raw material sectors, the trade & transport sector, and the non-food industrial sectors, the decline of imports in all agro-food sectors is over-proportional relative to the increase of import prices. This again replicates the developments immediately following the devaluation of the ruble after the financial crisis in mid–1998, when imports of food commodities declined most significantly.

The conclusion from this experiment is that the question, how negative terms-of-trade shock affect Russia's agro-food sectors is indeed an empirical issue. The response very much depends on the constellation in a given sector in the base period, on the trade parameters, and on the flexibility with which domestic production can respond to exogenously induced relative price changes.

Sensitivity analysis on Experiment 2 of Set 3

Experiment design. With Experiment 2 of the previous section (Set 3), we showed how the combination of two effects contributed to the particularly strong output decline in agriculture. However, what was still unsatisfactory in terms of validating the model was the fact that the simulation yielded a similarly strong output decline in small-scale agriculture. The specific response of the small-scale sector is plausible against the background of the model economy in that they indicate a much more moderate decline in the small-scale sector compared with the output decline in large-scale agriculture. However, the results of Experiment 2 were still not close enough to real world developments. At least in the course of the first transition years, an unprecedented increase in the share of agricultural output from small-scale agriculture was observed (see Table 1-2). This might indicate that the development of small-scale agriculture was also driven by other additional forces.

Changed model economy. We therefore simulated the same experiment (Experiment 2 of Set 3) by changing only the model economy in two consecutive steps. The results of these two additional experiments are reported in Table 5-8. In these two experiments, we continuously increased the flexibility of the model economy by changing the closure rules. In the first experiment, we changed the savings/investments closure by keeping the marginal propensity to save fixed and making investments flexible (Exp. 1). Additionally, in the second experiment, we made factor market adjustment more flexible by allowing for full intersectoral capital mobility, removing the disequilibrium condition of labor markets, and also making labor mobile between sectors (Exp. 2). In a variation of Experiment 2, we changed the foreign exchange market closure by fixing the exchange rate and ensuring a balanced foreign account via flexible foreign savings. As expected, this change in the foreign exchange closure did not alter the results of the experiment significantly because of which the respective results are not reported in detail.

Change in experiment design. However, as will be shown below, the alteration in closure rules yielded only partially satisfactory results. We therefore add yet another exogenous shock (Exp. 3): we increased TFP in small-scale agriculture by 10%. This last experiment should therefore not be interpreted as yet another sensitivity experiment but instead constitutes an additional variation in the original experiment design.

Experiment 1. The results of the experiment discussed in the following will be contrasted with those of Experiment 2 of Set 3 (see Table 5-6 and 5-7). By making investments flexible, the effects of this simulation on the Russian economy are altered significantly (see Exp. 1 in Table 5-8). On the one hand, GDP declines by more than one additional percentage point and employment by almost three additional percentage points. Here, the underlying assumption would be that investments could actually be withdrawn from the economy. In fact, the reduction of the rental rate in all sectors would, in a more flexible economy of this kind, result in a significant withdrawal of investments (ΔINV: -21.66%). Therefore, the exchange rate appreciates less and the decline of exports and imports is also less pronounced.

The sectoral effects of this experiment are also different. If investments are flexible, the burden of adjustment shifts towards those sectors whose production is used to a significant part to meet investment demand. Therefore, in this model economy, the experiment is particularly harmful for the construction and the machinery sector, both of which reduce output more significantly compared to Exp. 2 of the previous section (ΔX: -17.54 and -11.59%, respectively). It should be noted, that the argument could also be turned around. If the economy would experience an exogenously determined positive supply response and if investments would be flexible, the construction sector would show the most positive response. In fact, where investments within Russia have been – relatively speaking – flexible, such as in Moscow, the construction sector seems to have boomed in the second half of the 1990s.

In contrast, the adjustment in those sectors which produce consumption goods and not investment goods, such as the food industries and light manufacturing, is more moderate (ΔX: -9.45 and -8.29%, respectively). While the decline in large-scale agriculture is similar, the decline in small-scale agriculture is only about half as great as in the previous experiment. More importantly, the decline in small-scale production is much less pronounced than in large-scale agriculture, which increases the plausibility of the results.

Results of Experiment 2. Again, the results of this experiment are contrasted with those of Experiment 2 in the previous section (see Table 5-6 and 5-7). Here, the effects of the experiment on all macroeconomic indicators are much more moderate: GDP declines by only about 4%, the exchange rate appreciates by just 1.4%, and the reduction of exports and imports also remains limited at about 3%. These results are due to the longer run adjustment process simulated with this model economy. In a similar way to an economy which can be typified as neoclassical, all factor markets are fully cleared and any shock to which such an economy would be exposed would be compensated for by significant restructuring within the economy – which takes time. The restructuring would take place such that all resources remain fully employed and the economy would produce *on* the production possibility frontier (see Figure 2-6). This mechanism also reduces the negative effects on agro-food sectors and particularly on small-scale agriculture, which in this experiment reduces output at a much lower rate than large-scale agriculture. The fuel industry, which is the most export-oriented sector in the base period, increases output most significantly, which is another more realistic feature of this experiment.

Results of Experiment 3. It could still be argued that even the significant difference in output decline between large-scale and small-scale agriculture which resulted from the last experiment is not sufficient to replicate the unprecedented increase in Russia's small-scale agriculture in the transition period. It could also be questioned if the underlying closure rules which imply a fully flexible economy are plausible in the case of the Russian economy in 1994.

As we now have tested for various important types of 'model economies' that can typified with this modeling framework, we continue with the following question: what other changes might have induced not only a lower decline in small scale agriculture but even an increase in output in this sector? One option to answer this question with this modeling framework is to argue that there must in fact have been a positive shift of the supply function in this sector which was different from the shifts in all other sectors.

Such a *shift or improvement in total factor productivity* could be *explained by institutional changes* which altered the incentive system for the rural labor force. The most prominent features here are the *institutional changes* associated with farm management itself: with the collapse of the *kolkhozi* and *sovkhozi* system, the possibilities for the management of the large-scale farms to control the quality of labor declined further (Brooks and Lerman 1994). Shirking on the job reduced output on large-scale farms whereas, for instance, the non-taxation of income from small-scale farms (Krylatykh and Semyonova 1996) enhanced the motivation of the rural population to work more intensively and accurately on their small plots of land. While we do not elaborate further on the institutional factors which changed in this context and which might have affected the productivity of the small-scale sector, such a positive shift of this sector's supply curve seems to be a plausible explanation for the improved performance of this sector in the transition period.

The *results of Experiment 3* are along the roughly same lines as those of the last experiment, though slightly more moderate (Table 5-8). The notable exception is the relatively strong increase in production in small-scale agriculture, which is the direct effect of the positive shift in the sector's total factor productivity.

Conclusions. The discussion in the previous sections indicated that no single factor should be said to have caused the unprecedented output decline in Russia's agro-food sector during the first years of transition. In fact, the experiments showed that a wide range of factors which agricultural policy makers were frequently unable to control contributed to the output decline. The simulation on the transport sector indicated that the decline of TFP, not only in agriculture itself but also in other sectors, has very negative economy-wide effects. Due to the high importance of transportation for the agricultural sector in the Russian Federation, the underinvestments in this sector in the past decade are likely to have contributed to the negative development in agriculture. Kuhn (2000) extended this argument by showing that not only transportation costs but also transaction costs seem to increase over-proportionally with distance in Russia. Both arguments taken together hint at the fact that in a country of Russia's size inefficiencies in the trade and transport sector are likely to be very expensive, not only for the society as a whole but for agro-food sectors in particular.

Exchange rate effects

The effects of the first three sets of experiments on the exchange rate deserve further mention, as they are to some extent counterintuitive. Generally, the changes in the exchange rate in this model have to be interpreted firstly, against the data base and, secondly, against the closure rules characterizing our model economy.

Table 5-8 Macroeconomic and aggregate effects of sensitivity analysis for Experiment 2 of Set 3 (reduction in TFP in transportation sector by 20% and simultaneous reduction in domestic support to agriculture by 50%): with flexible investments (Exp. 1), fully flexible factor markets (Exp. 2), and by additionally imposing a 10% improvement in TFP onto the private subsidiary plot (LPH) sector (Exp. 3)

	Experiment 1[a])	Experiment 2	Experiment 3
Experiment design	20% reduction in TFP in transport sector and 50% reduction in domestic support to LAE	additional shock: 10% increase in	TFP in small-scale agriculture
Closure	flexible investments	flexible factor markets	as in Exp. 2
Macroeconomic results:		percentage change	
Real GDP	−8.88	−4.12	−3.69
Employment	−10.17	fixed	fixed
Exchange rate[c])	−5.59	−1.40	−0.98
Investments	−21.66	−4.55	−7.88
Exports	−5.82	−3.11	−3.09
Imports	−6.15	−3.31	−3.29
Sectoral output:			
Raw material sectors[d])	−6.28	−3.68	−3.63
Manufacturing	−10.33	−5.04	−4.95
Construction	−17.54	−6.92	−6.85
Food industries	−9.45	−7.20	−6.52
Agriculture, of which			
Large-scale agriculture	−10.74	−7.29	−6.34
Small-scale agriculture	−5.08	−2.59	3.72
Services	−8.22	−3.90	−3.81

Notes: a) Experiment design is the same as in Experiment 2 (Set 3) reported in Table 5-5 and 5-6. b) The closure in Experiment 1 is the Keynesian one used in all previous experiments with the only change reported here; all changes in closure rules in Experiments 1 and 2 are kept for the consecutive experiments as well. c) A positive (negative) change in the exchange rate indicates a devaluation (appreciation) of the ruble. d) See Table 5-3 for the sector aggregate, for which results are reported in the following.

Data base. The data in the base period reveals a trade surplus for the Russian Federation which is balanced by capital outflow, or as we call it, capital flight. This trade surplus would indicate that the ruble was undervalued rather than overvalued in the base period. Normally, it would be expected that a depreciation will increase the competitiveness of the export-oriented raw material sectors further. If foreign savings were flexible, such additional revenues from exports would directly increase the capital outflow.

Model economy. In our model economy, however, the exchange rate is flexible, while domestic savings abroad, i.e. foreign savings, are fixed in order to keep the trade balance fixed. Therefore, additional revenues from exports can not result in increased capital outflow. Hence, the real exchange rate needs to adjust to imbalances in the trade account which are due to exogenous shocks. Furthermore, in this model economy the sectoral wage rate is fixed (see explanation for Figure 2-7) which means the domestic price adjustment is also limited.

Against this background of the data base and the model economy, the first three sets of experiments which we discussed earlier have yielded two patterns of exchange rate responses (see Table 5-9):

- when terms-of-trade shocks are simulated, the real exchange rate depreciates and the economy contracts (Set1/Exp. 1; Set2/Exp. 1; Set3/Exp. 3);
- when the exogenous shock induces declining (increasing) TFP or a deteriorating (expanding) capital stock, the economy contracts (expands) and the exchange rate appreciates (depreciates; Set 1/Exp. 2; Set1/Exp. 3; Set 2/Exp. 2; Set2/Exp. 3; Set 3/Exp. 1; Set 3/Exp. 2).

These results are at first sight contradictory and therefore need to be explained further.

Adverse TOT shocks. The adverse TOT shocks we have simulated in the respective experiments were relatively moderate. The real depreciation that follows an adverse terms-of-trade shock would, in a neoclassical model, make *all* domestic goods and factors relatively less expensive, thereby inducing a restructuring within the economy (see Figure 2-5 in chapter 2). The new production location would be *on* the production possibility frontier.

In our model economy, however, the reaction to the adverse terms-of-trade shock is slightly different. Again, because of the fixed trade balance, the immediate response to the adverse terms-of-trade shock would be that the real exchange rate depreciates. However, one of the most important domestic prices – that of labor – is fixed, and capital is immobile between sectors. As wage rates can not adjust, the economy would release labor. Hence, the response to a terms-of-trade shock within our model is less pronounced than it would be in a neoclassical economy in so far as the reallocation of resources and, hence, the restructuring within the economy is limited. In contrast, the effects on GDP can be more significant than in the neoclassical economy.

In 1994, most Russian sectors relied on imported commodities, which became more expensive due to the terms-of-trade shock. Table 5-5 shows that imports in all tradable sectors are reduced. Furthermore, the release of labor reduces the income available in the economy, which means that private consumption declines, contributing to the overall contraction of the economy.

Changes in TFP or capital stock. These experiments exogenously induced an increase (decrease) in domestic supply. Under constant demand, this increase (decrease) in supply imposes downward (upward) pressure on the output prices in the respective sectors, and thereby inducing a positive (negative) income effect. Hence, as domestic prices decline (increase), more income would be available for buying imports. At the same time, the increase (reduction) in the productive capacity of the economy increases (reduces) the demand for imported inputs. However, the trade balance has to remain constant, which means that the exchange rate depreciates (appreciates), making imports more (less) expensive.

Table 5-9   Effects of experiments on GDP and on the exchange rate

Set	Exp.	Type of experiment	GDP change contraction: − expansion: +	Exchange rate Dep. or App.
1	1	Increase in import prices in selected industrial sectors (adverse ToT)	−	dep.
	2	Deterioration of capital stock in agriculture	−	app.
	3	Increase of TFP in agriculture	+	dep.
2	1	Increase in import prices in selected industrial sectors (adverse ToT)	−	dep.
	2	Deterioration of capital stock in agriculture	−	app.
	3	Increase of TFP in agriculture	+	dep.
3	1	Reduction of TFP in trade & transport sector	−	app.
	2	Reduction of TFP in trade & transport sector plus reduction in domestic support	−	app.
	3	Increase in import prices in all sectors (adverse ToT)	−	dep.

Validation of exchange rate response. Both types of responses are in line with what Russia has experienced in the 1990s. First, as mentioned above, in the course of the *crisis, adverse terms-of-trade* shocks forced the Russian ruble to depreciate. In contrast to our experiments, this adverse terms-of-trade shock was not experienced on the import side but on the export side. As a result of the declining international prices for oil and other energy exports, Russia's terms-of-trade deteriorated, which was basically a *spill-over effect* of the financial crisis (see SEROVA, VON BRAUN and WEHRHEIM 1999). In reality, this terms-of-trade shock was topped by a *contagious effect*, which added to the downward pressure on the ruble. Therefore, the devaluation of the ruble in the course of the crisis was much stronger than indicated in our experiments. In fact, with the model we can only show the terms-of-trade effect and, hence, the *spill-over effect*, but not the *contagious* effect.

Second, our simulations show that the *declining production efficiency* and the *deterioration of the capital stock* in large scale agriculture directly contributed to the appreciation of the exchange rate under a flexible exchange rate system. In fact, SEDIK, TRUEBLOOD and ARNADE (2000) found that the efficiency of Russian agriculture has worsened in the 1990s, as large-scale farms have moved further away from best national efficiency practice. At the same time, the ruble appreciated in 1994 strongly, contributing to the high competitiveness of imported food commodities in the 1990s (OECD 1999a).[4] While many factors might have driven the real appreciation of the ruble, our model results indicate that the declining efficiency in Russia's agriculture might have contributed to it and, hence, to making imported food commodities more competitive. It should be understood that under a flexible exchange rate system a real appreciation of the ruble could always happen again. In fact, in 2000 the OECD (2000b) stated that distortions in domestic prices that are associated with a real appreciation of the ruble could further jeopardize the economic stability of the Russian Federation. Hence, the model results highlight various risks that are associated with real appreciation of the ruble not only in the mid–1990s but to date.

Effects of Russia's WTO accession on agriculture

One of the most important economic changes in the early transition period was the process of *opening up* (see chapter 2). In the case of Russia, this process resulted in a far-reaching liberalization of the trade regime early on in the 1990s. This is also true for the agricultural trade regime, which had become fairly liberal by 1994, though protection increased in consecutive steps thereafter (see chapter 4). With the following simulations, we will analyze some of the options the Russian Federation has to further develop its trade regime. On the one hand, a decision *not to integrate* more into the global trading system and not to enter the WTO is likely to result in increased pressure of agricultural lobbyists and a subsequent rise in the level of protection of agriculture. On the other hand, *integration into the WTO* would mandate further tariff liberalization. Hence, we will carry out two sets of

---

[4]   In an empirical paper DYNNIKOVA (1999) presents results of cointegration analysis for the RF on the causality between the exchange rate, imports, and domestic production of various (food) commodities. The analysis suggests that the month-to-month real appreciation of the ruble between 1993 and 1997 which made imported goods cheaper, induced additional short-run increase in non-CIS meat imports which crowded out at least parts of the increase in domestic meat production.

experiments (Set 4 and Set 5), each addressing one of the two strategies mentioned and the respective effects on agriculture.

State of negotiations

Russia *applied for WTO membership* in 1993, but is still negotiating over the terms of its membership to this day. The expected effects of a potential WTO membership, and particularly of exposing the country's agro-food sector to the organization's rules on trade, are controversial within Russia. The Russian Federation became an observer to the GATT in 1990, submitted its application for WTO membership as early as June 1993, and the WTO Working Party, the council in which the various parties negotiate over the terms of accession, was established in the same month. In July 1994, the Memorandum on its Foreign Trade Regime was presented to the Working Party. In February 1998, the first tariff offer on goods was presented by the Russian delegation. By mid–1998, the information gathering process had reached an advanced stage, and bilateral negotiations on the terms of accession with major trading partners have been going on since then. Generally, the long duration of Russia's WTO accession process is typical for a large, formerly inside-oriented economy with many institutional features dating back to the centrally planned era and with an extreme overload of laws that often contradict each other and are enforced in various regions of the country rather differently (LANGHAMMER 2000: 25).

Controversial issues in the accession negotiations: agriculture sector topics[5]

In order to comply with the *WTO rules on agriculture*, which were set with the 'Agreement on Agriculture' at the end of the Uruguay Round, Russia would need to make *concessions in three major areas: market access, domestic support to agriculture, and export subsidies.* Currently, Russia's negotiating position with respect to agriculture is not to accept the low level of direct support that would be allowed if WTO rules were to be applied strictly. Instead, rather high levels of support and hence, protection, from world markets are demanded. More specifically, the following proposals are made:

Tariff offer and market access.   In early 1998, the Russian Federation made an offer on market access and on tariff bindings. The proposal offered a starting average bound level for products from the agro-food sector of 48%, an overall reduction of this bound level at the end of the implementation period to 37%, and a minimum reduction of 15% for each product group within the Harmonized System

---

[5]   The following section is based on information gathered by the author in the context of the SIAFT-TACIS-project in Moscow. During this assignment, the author was working for AFC consultants, Bonn/Germany, whose support is kindly acknowledged. Further information on the TACIS project, its objectives, results, and project reports on WTO accession of Russia and other members of the CIS can be obtained from http:// www.aris.ru/ in both English and Russian. Once at the home page of the Russian Ministry of Agriculture follow the links for SIAFT.

of tariffs. However, product-specific tariff offers were rather high, ranging between 15 and 68%. Hence, at the end of the implementation period, a final binding in the range of 10 and 53% would result, which would yield an average upper bound rate of 26%. Today, however, most import tariff rates for agro-food commodities range between 5% (for crop products) and 20% (for livestock and processed food products), which means that these higher bounds would not become effective and therefore would not result in any effective liberalization. Therefore, the trade partners in the WTO negotiations proposed to bind the upper level of average tariff rates at 14%, with a 36% reduction by the end of the implementation period, which would yield an average import tariff of approximately 9%. At the end of the year 2000, the Russian Ministry of Agriculture worked on a new tariff offer to take into account higher tariff cuts for those agricultural commodities which are not produced domestically, such as tropical fruits. It is expected that Russia might offer to reduce the upper bound average tariff rate to about 20%.

Domestic support to agriculture and base period. Russia proposed to use the period between 1989–91 as the base period for the calculation of the support level that would serve as a reference for defining the bound levels of support. During this period, the average level of support for agriculture (measured by the Aggregate Measure of Support or AMS) was US$ 84bn and thus relatively high even by OECD standards. It was argued by the Russian delegation that the agrarian sector during this period was still relatively stable and not yet distorted by the transition from plan to market. The member countries of the CAIRNS group[6] in particular oppose this view, as it distracts from the standard approach for WTO accession, which mandates that the three most recent years preceding the formal application are used as a base period. As a result of Russian opposition to this proposal, and because exceptions to this rule have been made in the case of other countries acceding the WTO in the past, a new offer on the base period covering the years 1989–95 was expected. Based on this period, the average maximum level of support would amount to US$42 billion.

Export subsidies. Russian agricultural policy makers believe that Russia will become a net exporter of at least some agricultural commodities in the future, and that export subsidies will be an adequate policy instrument to enhance the country's competitiveness in world food markets. Therefore, Russia applied for the right to grant relatively high levels of export subsidies to agriculture which, if implemented, would be comparable to the levels currently granted in the United States. The proposed level of export subsidies is based on the price difference between Russian sales to foreign countries and world market prices in 1989–90. As the former were lower at that time, the Ministry for Agriculture originally claimed a level of export subsidies of US$1.6 billion. In the meantime, Russia demands a

---

6   The Cairns Group of 15 agricultural exporting countries was formed in 1986 to influence agricultural negotiations within the WTO. Members of the Group are: Argentina, Australia, Brazil, Canada, Chile, Colombia, Fiji, Indonesia, Malaysia, New Zealand, Paraguay, the Philippines, South Africa, Thailand, and Uruguay.

level of export subsidies well below US$1 billion. Again, the base period used for the calculations of these subsidies is not the last three years prior to the application for WTO membership, but a different base period, the mid 1980s, chosen by the Russian negotiators.

Further liberalization.   In the preceding discussion, we outlined the three broad categories of policy instrument used by the WTO to classify the various agricultural policies in use. While the Russian Federation might attempt to raise the level of support with single policy instruments, the accession to the WTO might also involve support cuts across all products. Past experience of WTO accession negotiations has shown that acceding countries are likely to face demands for liberalizing the trade regime at least to some extent (MICHALOPOULOS 1998). Due to the importance agricultural trade liberalization had in the Uruguay Round of the GATT, agricultural trade policies of acceding countries have been reviewed with particular stringency in the 1990s. In fact, agricultural policies of a country seem to be under increasingly strict review the greater the agricultural export potential of the respective country. Thus, the Russian Federation should also consider the case for further liberalization of its trade regime.

Uniform tariff rates

Another issue that has been raised in the ongoing debate on developing Russia's tariff strategy and in the context of its WTO accession is the following: "Will uniform or escalating tariffs be the better choice?". A *uniform tariff* would be a 'flat tariff rate' that is applied to all products, independent of the commodity group and the degree of processing of the respective products. An *escalating tariff* schedule would fix lower tariff rates for raw materials and higher tariff rates for goods of the same commodity group that are processed further.

The World Bank estimated that custom tariffs for food products in the Russian Federation showed some degree of tariff escalation in 1997. While the average for both unprocessed and semi-processed food products was estimated to be about 10%, only the average for final food products was significantly higher (14.5%). Uniformity of Russian tariffs across all sectors was said to be lower than in many other countries, but there is evidence that various sectors are particularly protected with tariff peaks. At the same time, there seems to be little indication that those sectors of the Russian economy that were sheltered from foreign competition more than other sectors achieved significant productivity improvements (TARR 1998). The conclusion of TARR (1998) was therefore that uniform tariffs would serve the Russian Federation best. As discussed in chapter 4, the tariff schedule of the Russian Federation revealed, on the product level of our model and for 1994, moderate tariff variation.

Uniform tariffs and corruption.   Nevertheless, it is a strategic decision to use uniform tariffs or to keep the system of varying tariffs, which gives ample room not only for *rent seeking* but also for *corruption*. In fact, similar to most CIS countries, customs administration in Russia is not yet very efficient. The wide variety of trade policies, tariff rates and technical barriers to trade contribute to an often in-transparent trade system of the CIS countries. In this situation, uniform tariffs would have significant administrative advantages over strongly varying tariff rates: the more uniform tariffs were, the lower the need to classify imported goods in different custom groups would be. This would also reduce the discretion at which custom officials can classify the goods. If the category of a commodity is decisive for the tariff rate levied on the good and if the tariff rates are very different, an importer would have strong incentives to bribe the customs official so that the good is classified in the commodity group with the lowest tariff rate possible.[7] As corruption of customs officials is still a problem in the Russian Federation, uniform tariffs should be an efficient instrument to reduce corruption and the associated efficiency losses.

Furthermore, motivation to examine the issue of uniform tariffs stems from a Business Newsletter, released in October 2000 (BISNIS 2000). Here, it was stated that unifying Russia's customs duties was an important topic in the discussion of Russia's future customs policy. It was said, that the Government of Russia intended to develop a unified tariff level, reducing many tariff peaks to a level of 20%. In mid–2000, it was expected that the uniform tariff system would become effective in 2001.

Experiment design for effects of Russia not joining the WTO and increasing agricultural protection

To get an impression of the economy-wide effects of what would happen if Russia were not to join the WTO and instead increase the protection of agriculture, three different simulations were carried out. As the negotiating position of the Russian government with respect to the three policy instruments is still rather vague, we simulated *ad hoc* increases in the respective protection levels. As in the case of previous experiments, the results have to be interpreted against these specific protection levels and against the data in the reference situation. This will be particularly important with respect to agriculture, as exports from the Russian farm sector in the reference period have been very low. Furthermore, it was assumed that, from a political-economy point of view, the paths of these policies are independent of each other. Therefore, and to keep the results as transparent as possible, the *three types of agricultural support policies* were not simulated in a cumulative manner.[8]

The *first simulation* looks into the effects of *increasing the import tariffs in all agro-food sectors* which produce tradables, that is, all but the two small-scale farm sectors and the animal feed sector. To reach the approximate tariff levels proposed by the Russian delegation as upper tariff bounds, we doubled the import tariffs applied in 1994 in each of the agro-food sectors (see section 4.6). The *second simulation* looks into the effects of another Russian proposal in the accession

---

[7] In a theoretical paper, GATTI (1999) shows that in a situation in which corruption is pervasive "... setting trade tariffs at a uniform level eliminates officials' opportunities to extract rent".

[8] Because such a combination of policies might have additional general equilibrium effects, the estimation of the aggregate effects by simply totaling the effects of the three simulations was avoided.

negotiations by *introducing agricultural export subsidies* in the two large-scale sectors. However, no export subsidies were paid in the reference period and, therefore, we approximated the value of export subsidies at about 10% of the output price. The *third simulation* analyzes the effects of doubling the base-period level of *direct domestic support to large agricultural producers.*[9] Because the major share of these subsidies was granted in the mid–1990s as direct production support (OECD 1999a), we simulate this policy instrument via indirect taxes. In the discussion of the base data, it was shown that the two large-scale agricultural sectors were the only two sectors in the economy which did not pay any indirect taxes but received transfers from the government via negative indirect tax payments. Hence, the government had to finance these subsidies in 1994. The three experiments are therefore different in the choice of the policy instrument used to reach the objective of protecting and/or subsidizing Russia's agriculture and/or its domestic food industries and, thus, should induce different economic effects.

Simulation results: effects of Russia not joining the WTO and increasing agricultural protection

Table 5-10 summarizes the macroeconomic effects of the three simulations described above. The first two experiments, in which support to agriculture is granted with trade-distorting measures, result in a marginal overall decline of GDP. A result that is, at least at first sight, counterintuitive is obtained from the third simulation: here, the increase of direct domestic support to large-scale agriculture yields an increase of GDP by 0.44%. In the following, we will explain these differences in detail and show that they are in line with what would be expected against the background of our model economy.

Results of Experiment 1.   The rise in import tariffs in the agro-food sectors yields an increase in relative prices of imports over domestic supply in these sectors (P^M/P^D). The immediate effect is a decrease in total imports by 2.22%. As we fixed the trade balance in our model economy, total exports have to adjust accordingly and decrease by the same amount in foreign currency (–2.10%). This can only be achieved by an appreciation of the real exchange rate (–3.28%) which decreases the relative price of exports over domestic supply (P^E/P^D). Furthermore, in this experiment the economy contracts, however, the decline of real GDP and the increase in unemployment are – at 0.33% and 0.56%, respectively – marginal. In response, to this reduction in national income household savings decline by 1.66%, while the substantial government deficit is reduced by 5.62% because of higher import tariff revenues.

Another notable result of this experiment is the fact that *sectoral output restructures significantly*: the appreciation of the exchange rate particularly hurts the export-oriented raw material producing energy sectors as well as the manufacturing industries, which both reduce output by about 1%. In contrast, the domestic

---

[9]   The opposite shock – a 50% reduction of direct domestic support – was also part of Experiment 2 in Set 3 (see p. 131).

non-tradable sectors are better off and either reduce output by very little, for example the service sectors (ΔX: –0.11%), or even grow, for example the construction sector (ΔX: 0.56%).

The experiment has *mixed effects on the agro-food sectors*. Due to the increase of import tariffs and, consequently, increased relative import prices, the results are as expected: domestic demand shifts from imported commodities to domestically produced goods, the degree of which is determined by the magnitude of the change in relative prices (P^M/P^D) and the sector-specific CES value (see chapter 4). On the one hand, the increase of import tariffs increases the price of imported food commodities used as intermediates and for final consumption. All agro-food sectors producing tradables reduce imports significantly (see Table 5-11). On the other hand, this effect is offset by the sizable appreciation of the exchange rate, which favors domestically produced products. As a result of these diverging forces, output in the food industries decreases slightly (ΔX: –0.14%).

An output decline of similar magnitude is experienced by small-scale agriculture (ΔX: –0.18%). The reduction of income – and hence the decline in household expenditures – hurts this sector most. In contrast, large-scale agriculture increases output, but only by 0.81%. This increase is the immediate effect of increasing border protection in this sector via higher import tariffs, which means that imports in large-scale crop and livestock production are reduced significantly (ΔM: –16.64% and –10.90%, respectively).

These results indicate that, even though the effects of preferential treatment of the agro-food sector on GDP might be limited, they induce significant restructuring and hurt other sectors substantially. At the same time, the experiment highlights that rising import tariffs for agricultural commodities represent a two-edged sword in a situation of high import dependence. It proved harmful to the food processing industries, which are hurt by the higher prices for their intermediates.

Sensitivity analysis.   The increase in import tariffs in the agro-food sectors is likely to persist in the long run once it has been implemented. Therefore, we simulated the same experiment (Exp. 1 of Set 4) by increasing the trade elasticities continuously. Starting with the CET and CES values discussed in chapter 4, we increased both elasticities in each experiment by an absolute value of one (Exp. 1 to 7). This yielded CES elasticities of 6.6 in the raw material sectors, of 8 and 7.5 in light manufacturing, in the food industries, and in the large scale agricultural sectors. The CET values increased up to a value of 8.9 in the fuel industry, and ranged in all other sectors in the last experiment between 6.5 and 8.

Figure 5-2 shows the percentage changes for some macroeconomic indicators. The doubling of import tariffs in the agro-food sectors increases the import price of the respective goods. If trade elasticities are increased the substitution between domestic and imported commodities should become more distinct. This is shown in Figure 5-2, in which the decline of imports is stronger the greater the elasticities gets (lowest in Exp. 1 highest in Exp. 7). To keep the trade balance fixed, the adjustment of the exchange rate needs to be more significant because the prices for exports do not change. The stronger the appreciation of the exchange rate, the bigger is the decline in exports. This, in turn, implies that with increasing trade elasticities the model yields stronger restructuring towards domestic production because of which the decline in GDP and consecutively in employment becomes less pronounced.

Table 5-10 **Macroeconomic and aggregated effects of three experiments (Set 4) increasing support to Russia's agro-food sector with three different policy instruments: doubling of import tariffs in all agro-food sectors (Exp. 1), introducing a 10% export subsidy for production of large agricultural enterprises (Exp. 2), and doubling of direct domestic support to large agricultural enterprises (Exp. 3)**

Experiment design Macroeconomic results	Experiment 1:[a] increase in import tariffs in agro-food sectors	Experiment 2:[b] export subsidies for large-scale agric. percentage change	Experiment 3:[c] domestic support for large-scale agriculture
Real GDP	−0.33	−0.22	0.44
Exchange rate[d]	−3.28	−1.71	0.83
Employment	−0.56	0.34	1.11
Government savings[e]	−5.62	7.97	4.85
Household savings	−1.66	1.84	1.27
Exports	−2.10	0.34	0.46
Imports	−2.22	0.21	0.51
Sectoral output:			
Material production	−1.05	−0.55	0.31
Manufacturing	−1.03	−0.62	0.57
Construction	0.56	0.26	−0.02
Food industries	−0.14	−0.22	2.94
Agriculture, of which			
Large-scale agriculture	0.81	8.90	8.25
Small-scale agriculture	−0.18	−0.20	1.91
Services	−0.11	−0.23	−0.30

Notes: a) Doubling of *import tariffs* in all agro-food sectors. b) Introduction of 10% export subsidy for exportable produced by the two large-scale agricultural sectors. c) Doubling of direct *domestic support* to both large-scale agricultural sectors. d) A positive (negative) change in the exchange rate indicates a devaluation (appreciation) of the ruble. e) Government savings in the base period were negative, i.e. the government ran a deficit, which meant a negative (positive) change of government savings, reducing (increasing) the deficit.

Table 5-11   Sectoral effects of three simulations (Set 4) on output, output and value added prices, exports and imports

Sector	Experiment 1[a] $X^{[d]}$	PV	M	Experiment 2[b] X	PX	E	Experiment 3[c] X	PV	E
1 EP	−0.45	−2.31	0.62	−0.33	−1.41	−0.79	0.42	2.22	−0.69
2 FI	−0.61	−3.61	−0.07	−0.32	−1.67	−0.33	0.15	0.89	0.10
3 MI	−1.50	−4.66	−0.03	−0.74	−1.31	−1.30	0.28	0.88	0.58
4 CI	−1.88	−3.52	0.26	−0.89	−1.38	−1.30	0.41	0.77	0.29
5 MB	−1.05	−1.49	2.54	−0.52	−0.79	−1.66	−0.06	−0.08	0.78
6 WI	−1.35	−1.73	0.69	−0.74	−0.85	−2.05	0.56	0.72	0.64
7 LM	−1.01	−1.68	2.95	−0.77	−1.24	−1.43	1.61	2.71	0.50
8 CO	0.56	0.32	2.24	0.26	−0.09	-	−0.02	−0.01	-
9 SR	0.42	1.65	−7.54	−0.34	−1.68	−0.34	2.36	9.68	2.76
10 FM	−0.57	−0.91	−6.83	0.09	−1.78	0.13	6.20	10.10	8.78
11 MP	0.47	2.19	−8.93	−0.09	−1.83	−0.03	2.61	12.65	3.03
12 DP	0.49	1.53	−8.54	−0.11	−1.75	−0.08	1.95	6.14	2.04
13 OF	−0.59	−1.85	−7.11	−0.59	−1.47	−0.78	1.12	3.81	1.03
14 AF	0.85	2.20	-	0.98	−0.70	-	−13.2	−30.38	-
15 CP	1.26	1.37	−16.64	9.71	5.63	245.72	4.04	4.40	11.76
16 LP	0.19	0.09	−10.90	10.98	1.01	218.62	22.64	10.15	33.16
17 LPH	−0.18	−0.28	-	−0.21	−0.46	-	1.92	2.96	-
18 PF	−0.12	−0.20	-	0.05	−0.26	-	1.75	2.79	-
19 TT	−0.26	−0.48	1.27	−0.10	−0.27	−1.11	0.27	0.51	0.48
20 SE	−0.02	−0.01	-	−0.32	−0.34	−0.06	−0.70	0.46	-

Notes: a) Doubling of *import tariffs* in all agro-food sectors. b) Introduction of a 10% *export subsidy* in exports of large agricultural enterprises. c) Doubling of direct *domestic support* to large agricultural enterprises. d) X = domestic output quantity in each sector; PX = price of domestic output; PD = domestic sales price for sectoral good; PV = price of value added. M = Imports. Names of Sectors: 1 Electric power; 2 Fuel industry; 3 Metal industry; 4 Chemicals industry; 5 Machine building; 6 Wood industry; 7 Light manufacturing; 8 Construction; 9 Sugar refining; 10 Flour milling; 11 Meat processing; 12 Dairy processing; 13 Other food; 14 Animal feed; 15 Large-scale farms: crop production; 16 Large-scale farms: livestock production; 17: Small-scale farms: private subsidiary plots (LPH); 18 Small-scale farms: private farmers; 19 Trade & Transport; 20 Services.

This is shown in Figure 5-3 which presents the effects of increases in the trade elasticities on the output change for various sector aggregates. The general message from this graphical presentation is, that the greater the trade elasticities, the more positive is the effect for the winners of this simulation, namely the large-scale agricultural sectors and the agro-food industries. The mirror-image of the increasing output expansion in the agro-food industries is an increasing decline in output of the loosers of this simulation, namely the export-oriented sectors. The decline in these sectors is less pronounced than the increase in the agro-food sectors due to the shares of the respective sectors in total trade.

These results underline that the inflexibility of our short run, Keynesian model economy contributes to the unemployment effects due to limited scope for restructuring. Furthermore, this sensitivity analysis indicates that the response of the model to increases in the value of trade parameters is theoretically correct and the extent of the response is plausible.

Results of Experiment 2. The introduction of export subsidies has somewhat similar results on the economy to the increase in import tariffs. GDP declines only slightly by –0.2%, and the exchange rate appreciates, but to a lesser extent than in Experiment 1 (by 1.71%). The 'taxpayer' has to finance the export subsidies granted to agriculture. The primary source for these subsidies is the government deficit, which increases significantly by almost 8%. In order to maintain the savings-investment balance, one of the savings components has to adjust in the

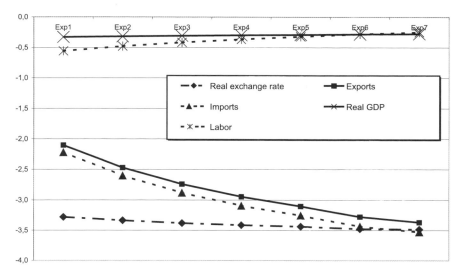

Figure 5-2 **Effects on macroeconomic indicators (percentage change) of continuously increasing trade elasticities (CES and CET) in seven experiments, while simulating an increase in import tariffs for agriculture in all experiments**

opposite direction. In our model, where foreign savings are fixed, household savings is the only component that is left to adjust. Indeed, it increases by 1.8% despite the fact that GDP declines. To keep the savings-investment identity balanced, the absolute change in household savings is about the same as that in government savings only with opposite signs.

While sectoral exports decline in most other sectors, most notably in the non-food industrial sector, they increase significantly only in large-scale agriculture (see Table 5-11). The additional demand for the good produced by large-scale agriculture for exports induced an increase in the domestic output price for the composite good of both sectors (*ΔPX:* 5.63% an1.01%, respectively) while this price declined in all other sectors.

The *diverging effects on exports* are due to the fact that only the respective two sectors benefit from the introduction of export subsidies, while all food industries but the flour milling industry are effectively worse off because of the appreciation of the exchange rate, which hurts the import-dependent food industries. The flour milling industry benefits from the increased demand in the livestock and animal feed sectors. In contrast, the construction sector, which effectively is a non-tradable sector, becomes better off and also increases output slightly.

The percentage change of exports in both large-scale agricultural sectors is tremendous because the initial value of exports was very low compared to the high amount of absolute export subsidies introduced by the experiment. However, the absolute change in exports in both sectors accounts for just about half of the output increase in both large-scale agricultural sectors. This *positive multiplier effect* is

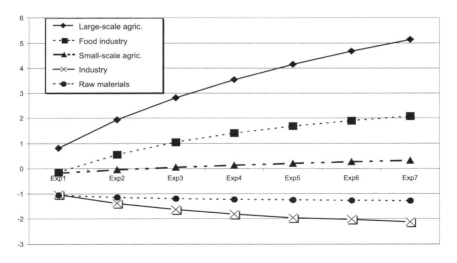

Figure 5-3   Effects on sectoral output (percentage change) of continuously increasing trade elasticities (CES and CET) in seven experiments, while simulating an increase in import tariffs for agriculture in all experiments

basically due to the labor market closure, which allows for hiring additional labor at the fixed nominal wage rate from the unlimited labor pool. In fact, the economy-wide positive effect on employment of this simulation is mainly driven by an increase in labor of 21.3 and 16.6% in the large-scale crop and livestock sectors, respectively, which is far greater than the respective output effect. This is so because capital is fixed and labor is the only primary production factor of which more can be employed in the production process to expand output.

The results highlight that, in this model economy, the subsidization of agriculture via export subsidies is harmful to the rest of the economy as a result of two effects: on the one hand, the taxpayer has to finance the subsidy and the pressure on the government budget is increased; on the other hand, the need to maintain macroeconomic balances results in a domestic bias, which in turn results in substantial restructuring, effectively discriminating against all other sectors. Hence, the model shows what a partial equilibrium model would hide: the support of one sector can only come at the expense of other sectors.

Results of Experiment 3. In this experiment, Russia's agriculture receives preferential treatment via direct support programs financed directly from taxpayers, and also via the government budget. As mentioned above, the most striking fact is that GDP actually increases by 0.44% in this experiment. There are two major explanations for this difference: first, the exchange rate effect is more moderate and, second, the output growth in large-scale agriculture can feed into growth in the domestic food industries because of positive forward linkages. The adjustment of the exchange rate – a depreciation of 0.83% – is lower than in both other experiments, which again mandates more moderate adjustments and restructuring in other markets. In fact, the results highlight that the increase in domestic support is a more direct policy in that it feeds immediately into an increase of domestic output in domestic large-scale agriculture, which can be used to meet domestic demand. The effects on the government budget and on the savings-investment identity are similar to those in Experiment 2, as the direct income support is financed via the same mechanism as the export subsidy: an increase of the government budget deficit and household savings.

The positive supply response in the two large-scale farm sectors, and here particularly in the livestock sector (see Table 5-11), results in a decline in sectoral output prices in these two sectors. As a result of the strong forward linkages to the food industries and the decline in large scale agriculture's output prices, the value added price in these sectors increases.[10] Therefore, the food industries also increase output, part of which is supplied to foreign markets. As there is growth not only in agriculture but also in the food industries, additional income is available in the economy, and those sectors with high household expenditure shares benefit from this: in addition to the agro-food sectors, this effect is particularly positive for light manufacturing, which increases output by more than 1.6%. Furthermore, it is obvious that the export subsidy induces substantial relative export growth in both

---

[10] The changes in sectoral output prices are not reported in Table 5-11. In the large-scale and crop-producing farm sector, this price declines by 6.5%, while it declines in the large-scale and livestock-producing sector by 9.1%.

large-scale sectors, even though there is a low absolute level of exports (ΔE: 11.8 and 33.2% in the crop and livestock sector, respectively). Hence, when arguing in favor of export subsidies in the course of WTO accession negotiations, Russian policy-makers should bear in mind that any export subsidy is effectively associated with a transfer of an economic rent from domestic taxpayers to foreign consumers.

Concluding comments on the experiments in this section. All policies simulated with the three experiments discussed in this section have a similar objective: the large-scale farm sector and/or the whole agro-food sector is supposed to benefit from additional sector support and/or increased protection from world markets. Even though the respective policies differ in the degree of support granted to agriculture, the results indicate important differences. Particularly notable is the effect that policies which distort trade have stronger macroeconomic repercussions because – under a flexible exchange rate system – they induce direct adjustments of the exchange rate and thereby more significant restructuring. The effects for the agro-food industries are quite mixed because various forces were affecting these sectors in our model economy: first, the adjustment to a change in the exchange rate; second, the backward linkages to agriculture as an important source of intermediates for the food industries, and third, the adaptation of domestic household demand. Furthermore, it is evident that, even though the third experiment was beneficial for the economy as a whole, the constrained resources that are available to the Russian government have reduced the degrees of freedom in designing agricultural policies associated with direct costs for the government. As long as the fiscal situation of the Russian Federation does not improve markedly, it is much more likely that indirect support policies such as increasing import tariffs for the agro-food sectors will remain the first choice of policy-makers. However, the results of Experiment 1 indicate that such a policy would be particularly harmful to consumers in Russia because of the high share of their income they must spend on food.

Experiment design for effects of further liberalizing Russia's trade regime

Against the discussion earlier, we will conduct another three experiments, each of which relates to specific trade strategies currently under discussion in the context of WTO accession of the Russian Federation. To make the point, we will simulate extreme cases with the three next experiments. The *first experiment* will look into the effects of a *full liberalization of the agricultural sector*. Therefore, this experiment is basically a combination of Experiments 1 and 3 from the previous section, but with opposite signs. All import tariffs in the agro-food sectors are removed and the direct support granted to large-scale agriculture is also removed. With the *second experiment*, we ask what the economic effects of *fully abolishing import tariffs in all sectors* would be. In fact, this strategy has been pursued by Estonia and seems to have contributed to the success story of this Baltic "Tiger State". With the *third experiment* in this section, we will look into the effects of modifying the structure of Russia's import tariffs in such a way that a *uniform tariff rate* is introduced. This policy experiment was conducted by keeping tariff revenues at the same level as in the base run. Distributing the tariff revenues over all imports yields a *flat tariff rate of 10%*. However, it is expected that this redistribution of tariffs will not yield significant effects because of the relatively moderate tariffs variation in the base run version of the model. We will therefore also conduct a complementary scenario by simulating a uniform flat rate of 5%, effectively cutting the tariff revenues in half.

Simulation result: effects of further liberalizing Russia's trade regime

Experiment 1. Against the background of our model economy and the data of the base period, the removal of all support granted to the agro-food sectors yields a decline in GDP of 0.2% (see Table 5-12). The removal of import tariffs makes imports less expensive and domestic products relatively more expensive. To keep the trade balance at its predetermined level the real exchange rate depreciates because of which exports become more competitive and increase at the same magnitude as imports. There is also a negative employment effect due to the negative GDP effects and which is again slightly greater than the latter because labor demand is the only primary production factor left to adjust. The government deficit increases further, despite the fact that government revenues from indirect taxes are reduced because no more payments are made to large-scale agriculture. However, the removal of all tariffs reduces the tariff revenues from agro-food imports by almost 40%, which makes the government worse off.

The sectoral effects indicate the expected response (Table 5-13): large-scale agriculture, which suffers from the removal of both tariff protection and direct domestic support, reduce output by more than 10%. Food industries and small-scale agriculture also reduce output, though for different reasons. The food industries reduce output because of the reduction in import protection which reduces the price of imports, but also because of the increasing output prices, at least for the good produced by the large-scale crop sector. All food industries, with the exception of the flour milling sector, increase imports. In response, the value added price declines not only in the two large-scale farm sectors but also in all food industries (see Table 5-13). The negative effects in the small-scale agricultural sectors are again the result of the strong forward and backward linkages with large-scale agriculture: the private subsidiary plots in particular 'buy' the good from the large-scale sector as an intermediate, which at least in the case of the large-scale crop sector became more expensive because of the reduction in domestic output, contributing to the decline in the value added price of this sector.

In contrast, the non-food industrial sectors and the service sector benefit from the liberalization of the agro-food sector. Both sector aggregates increase output slightly. However, as long as the flexibility within the model economy is limited, the full potential of liberalization can not be exploited: the scarce production factor capital can not be employed by those sectors which became relatively more competitive. Therefore, the restructuring can not compensate for the reduction in output in the agro-food sectors.

This is particularly true for the additional resources released from a reduction in direct domestic support to agriculture. This was shown by an additional simulation in which we only removed all import tariffs but maintained the direct domestic support to large-scale agriculture. While the depreciation in that experiment was – at almost 3% – much higher, real GDP actually increased by 0.3%. The reduction of output in the food industries was almost nil, that of large scale agriculture just about 0.6%, and small-scale agriculture increased output marginally by 0.1%.

Table 5-12　Macroeconomic and aggregate effects of three experiments (Set 5): full liberalization of agro-food sectors (Exp. 1), full liberalization of all sectors (Exp. 2), and introduction of uniform import tariff rate of 10% (Exp. 3)

Experiment design	Experiment 1[a] Full liberalization of agriculture	Experiment 2[b] Removal of import tariffs in all sectors of the economy percentage change	Experiment 3[c] Uniform tariff rate of 10% Tariff revenues constant
Macroeconomic results:			
Real GDP	−0.18	0.93	0.00
Exchange rate[d]	1.73	6.16	0.05
Employment	−0.39	1.93	−0.04
Government savings[e]	1.83	13.80	−0.17
Household savings	0.60	3.94	−0.04
Exports	1.16	5.03	−0.07
Imports	1.26	5.42	0.02
Sectoral output:			
Raw material sectors	0.48	2.45	−0.02
Manufacturing	0.08	0.61	−0.04
Construction	−0.49	0.09	−0.10
Food industries	−3.59	0.83	0.03
Agriculture, of which			
Large-scale agriculture	−10.39	0.52	−0.04
Small-scale agriculture	−2.39	0.78	−0.05
Services	0.75	0.59	−0.00

Notes:　a) Removal of import tariffs and direct domestic support granted to domestic agriculture (Exp. 1). b) Removal of import tariffs in all sectors (Exp. 2). c) Introduction of a uniform tariff (Exp. 3). d) A positive (negative) change in the exchange rate indicates a devaluation (appreciation) of the ruble. e) Government savings in the base period were negative, i.e. the government ran a deficit, with the result that a negative (positive) change of government savings means that the deficit is reduced (increased).

Additional simulations. To prove this argument, we carried out the same experiments using a fully flexible model economy in which labor markets are in equilibrium in as far as total labor supply is fixed and labor is mobile between sectors. Furthermore, capital and investments can adapt. First, we simulated the same shock as in Experiment 1 of Set 5, i.e. we removed all import tariffs in the agro-food sector and the direct domestic support granted in the base period to agriculture (Exp. 1B). Second, we simulated only a liberalization of agro-food sectors' trade regime while maintaining the level of direct domestic support to agriculture (Exp. 1C).

Results of additional experiments (not reported in any Table). Both experiments (Exp. 1B and Exp. 1C) yield a slight increase in GDP (0.13% and 0.08%, respectively). The marginal degree of GDP increase is again due to the fact that the underlying model now represents a neoclassical economy in which any exogenous shock results in a new equilibrium and a production point on the production possibility frontier. At the same time, this move along the production possibility frontier can be associated with significant restructuring because production sectors can be shifted into those sectors which are relatively more competitive. In both experiments, the agro-food sectors reduce output while the raw material producing industries are the ones which expand most (by 3.19% and 2.00%, respectively). While large-scale agriculture reduced output in Exp. 1B by almost 10%, the output reduction in this sector in Exp. 1C is more moderate – less than 2%. The results of these additional simulations should underline that the theoretical argument in favor of free trade is also valid in the case of the Russian economy. At the same time, these simulations highlight that the reduced flexibility within the Russian economy can in fact prevent that the full potential of trade liberalization is exploited.

Results for Experiment 2 of Set 5. The results of the second experiment in this section are complementary to those of Experiment 1 in that here the import tariffs of all sectors are completely removed (see Table 5-12 and 5-13). Against the background of our Keynesian model economy, a sizable increase in GDP and an even greater increase in employment indicate the overall positive implications of liberalization of Russia's trade regime.

However, given the short-term character of the simulation, this relatively strong exogenous shock would also yield other significant macroeconomic effects. On the one hand, one of the most immediate effects of the removal of all trade tariffs would be a restructuring of trade: as imports become relatively less expensive, the real exchange rate would depreciate by more than 6%. In order to keep the foreign trade balance in equilibrium, this would induce nominal exports to grow at a similar rate as imports. On the other hand, the liberalization would have strong effects on the government budget and the savings-investments identity. As the government would lose all revenues from import tariffs (–12.3 tR), it would need to finance some of its expenditures by increasing its budget deficit by 13.8%. This, in turn, has an effect on the savings-investment identity. In order to keep this identity in equilibrium, again the only savings component which can counterbalance the increased pressure on the government savings is household savings, which increase by exactly the same absolute amount by which the government savings decline (ΔGOVSAV: 13.80% and ΔHHSAV: 3.94%, in relative terms).

Table 5-13 **Sectoral effects of three simulations (Set 5) on output, output and value added price and imports**

Sector[b]	Experiment 1 Full liberalization of agriculture			Experiment 2 Removal of import tariffs in all sectors of the economy			Experiment 3 Introduction of uniform tariff rate of 10%; tariff revenues constant		
	X[a]	PX	PV	X	PX	M	X	PX	PV
1 EP	−0.17	−0.01	−0.86	1.14	4.20	2.69	−0.01	0.03	−0.06
2 FI	0.32	1.61	1.92	1.18	5.75	5.08	0.01	0.06	0.01
3 MI	0.85	1.33	2.71	3.42	3.96	5.17	−0.04	0.11	−0.13
4 CI	1.05	1.06	2.00	5.08	3.89	3.74	−0.12	0.15	−0.23
5 MB	1.03	1.05	1.47	0.78	0.26	3.87	0.17	0.24	0.24
6 WI	0.45	0.35	0.57	2.91	2.21	4.17	0.03	0.02	0.04
7 LM	−1.49	−0.64	−2.47	0.31	−0.76	10.47	−0.38	−0.71	−0.63
8 CO	−0.49	−0.05	−0.28	0.10	0.34	-	−0.10	−0.05	−0.06
9 SR	−4.26	−2.32	−15.82	−0.36	−0.42	8.35	−0.56	−0.44	−2.19
10 FM	−6.67	3.73	−10.45	1.47	1.17	−2.12	0.56	1.17	0.90
11 MP	−4.05	−1.02	−17.43	0.03	−0.01	9.77	−0.37	−0.58	−1.68
12 DP	−3.13	−2.35	−9.37	−0.02	−0.15	9.37	−0.34	−0.72	−1.05
13 OF	−1.05	−0.98	−3.24	1.42	1.98	1.09	0.14	0.64	0.43
14 AF	15.49	15.40	44.55	−0.60	−0.09	-	−0.11	0.36	−0.27
15 CP	−4.46	7.89	−4.85	1.04	0.92	0.35	0.74	0.51	0.80
16 LP	−28.90	−2.36	−3.68	0.25	1.09	11.96	−0.05	−0.05	−0.51
17 LPH	−2.41	−1.32	−3.39	0.78	0.85	-	−0.04	−0.03	−0.08
18 PF	−2.15	9.88	−14.92	0.71	0.27	-	−1.07	−0.15	−0.06
19 TT	0.02	0.06	0.04	0.83	1.42	2.04	−0.01	−0.03	−0.02
20 SE	1.26	0.67	0.82	0.42	0.33	-	0.00	0.00	0.00

Notes: a) X = domestic output quantity; PX = price of domestic output; PV = price of value added. M = Imports. b) Names of Sectors: 1 Electric power; 2 Fuel industry; 3 Metal industry; 4 Chemical industry; 5 Machine building; 6 Wood industry; 7 Light manufacturing; 8 Construction; 9 Sugar refinery; 10 Flour milling; 11 Meat processing; 12 Dairy processing; 13 Other food; 14 Animal feed; 15 Large scale farms: crop production; 16 Large scale farms: Livestock production; 17: Small scale farms: private subsidiary plots (LPH); 18 Small scale farms: Private farmers; 19 Trade & Transport; 20 Services.

In comparison to all previous simulations, the sectoral results of this experiment show an overall optimistic effect of this far-reaching liberalization. Again, it is the raw material producing sectors which benefit most from the liberalization, and which leads to an output increase in this sector of 2.5%. At the same time, the agro-food sectors increase output moderately due to the fact that imports become less expensive. All but one sector (flour milling) increase imports, which effectively lowers the production costs in these sectors. While the effects are most moderate in the non-traded sectors, the small-scale agricultural sectors grow on average even more than the two large-scale sectors together. This is due to the increase of income in the economy, which is used for increasing the expenditures on the good sold by small-scale agriculture and secondly to a reduction in the price for intermediates for both sectors, which means that the value added price increases in both sectors more than in the large-scale sectors.

Furthermore, it should be noted that the positive output response in the raw material producing sector would have been even greater and the sectoral pattern of this experiment's output response would have been more distinct if we had simulated not only a liberalization on the import but also on the export side. The abolition of all export taxes levied on the industrial sectors in the base period would have made these sectors even more competitive compared to other sectors of the Russian economy.

This simulation indicates that, even under the conditions of a relatively inflexible Russian economy, a liberalization strategy would be superior to one of higher protection. If the domestic production potential is fully released, part of the labor force, which in the first transition decade was still employed in the unofficial economy, could be shifted to more productive employment opportunities in the official economy.

Results for Experiment 3. This last experiment looks into the effects of introducing a *flat import tariff rate* for all tradable sectors of the Russian economy. The results of this simulation indicate that the macroeconomic effects of introducing a flat rate of 10% are almost nil. This is so because the variation of import tariffs in the base period was not very strong, and because tariff revenues were kept constant. Sectoral restructuring is also negligible. This result, however, could be interpreted in favor of such a flat rate if it is complemented by the argument that has been proven by GATTI (1999): introducing a uniform import tariff rate eliminates officials' opportunities to show corrupt behavior, thereby increasing the welfare of the society. In order to show that – on top of this argument – there are additional benefits from using a flat rate, we extended this experiment. In Exp. 3B we did not keep tariff revenues constant but instead introduced a uniform import tariff rate of only 5%, thus effectively liberalizing Russia's trade regime. The macroeconomic and sectoral results of this simulation are proportional to those of Experiment 2 in this section because this experiment is basically identical to a 50% liberalization of the country's import trade regime. The experiment yields an overall increase in GDP of 0.5% and an increase in employment of 0.9%. All sectors achieve a positive output response, which is most distinct in the raw material producing sectors. Hence, these results highlight that if a flat import tariff rate would be chosen, one that effectively liberalizes Russia's import trade regime would be superior.

Chapter 6
Summary, Policy and Research Conclusions

Summary

The major objective of this study was to develop an applied computable general equilibrium model of Russia's economy which pays special attention to agriculture and the food-industries and which reflects some of the most important features of Russia's economy in transition. The theoretical expositions in chapter 2 indicated that the development of Russia's agro-food sector is dependent on *sector-specific factors* as well as on *sector-neutral events*. The use of a computable general equilibrium model for Russia allows both, a focus on linkages among various agricultural sub-sectors and the food industries in an economy-wide context.

In the introductory chapter, some *stylized facts* about the general economic development of the Russian economy in the 1990s as well as on the changing role of agriculture in Russia's transition period were presented. We showed that the contribution of agriculture to the economy declined significantly in the 1990s. In that period, primary agricultural production experienced a significant output decline and restructuring of production took place in that the output share of the former large-scale farms declined to below 50%, while the share of small-scale producers (private farmers and producers on household plots) increased tremendously. In chapter two, *theoretical arguments* on the country-specific transition were given. We argued that the general economic situation, which is characterized by a high level of risk, has influenced the poor output performance in both the Russian economy as a whole and in the agricultural sector. In fact, some *economic inflexibilities* in the Russian economy are the norm rather than the exception. We showed that these market imperfections have implications for a general equilibrium analysis and discussed the differences between a fully flexible neoclassical general equilibrium analysis and a *structural general equilibrium model*. In the structural model, we proposed to take *unemployment* explicitly as one typical feature of the transition process into account. Therefore aggregate demand has a much higher relevance. Furthermore, the structural model can highlight the importance of reducing economy-wide market imperfections.

Taking the model requirements outlined in chapter 2 into account, we described in chapter 3 the *theoretical structure of the one-country-CGE model* and the various mandatory steps in developing such a model. At the end of this chapter, we described how the standard model has been modified in such a way that it incorporates some important structural features of the Russian economy so that it represents a structural model. Chapter 4 presented the *data base of the model* which is also crucial for developing a model of the Russian economy that is as realistic as possible. We started out with a *consistent data set for 1990*, deducted from an IOT for the Russian Federation as published by the World Bank, and updated it to 1994 using secondary data. The updated data base distinguishes within primary agricultural production between *two large-scale and two small-scale farm sectors.* In chapter 5, we used both model versions, that for 1990 and that for 1994, to simulate various sector-specific events (economic and policy changes) as well as sector-neutral shocks. In the following, we will discuss the implications of the simulations in more detail. All of the simulations have to be interpreted cautiously and against the background of the specific model-economy as discussed in chapters three and four.

While *applied general equilibrium analysis* has become a commonly used tool in policy analysis for developing countries its application to transition economies is still in its infancy. Against this background, this study provided a *first theoretically and empirically consistent application of the framework to the Russian Federation.* The standard neoclassical CGE model was modified in such a way as to incorporate some important transition-specific features. Particularly allowing for unemployment of production factors and, thereby, *introducing a notion of disequilibrium* brings the model closer to reality. Furthermore, in most other CGE studies of transition economies, agriculture, if it has been disaggregated at all, was split up according to various production branches. In our analysis, the *split was performed according to the most important farm structures.* In doing so structural features of different farm types could be represented. Most importantly, the disaggregation of primary agriculture indicated, even though in a very stylized manner, how the dualism between the former collective, large-scale agricultural enterprises and the small-scale agricultural producers can be accommodated in a CGE framework. The *subsistence character of small-scale agriculture was taken into account* by high shares of inputs from the sector itself, a high share of labor in value added, and high shares of direct sales to final consumers instead of to the food industries. While this representation provided important insights, it can be extended further in future versions of the model by incorporating further characteristics of the different agricultural producers.

Research conclusions

Results related to research question 1: potentials and limitations of the general equilibrium framework

The objective of constructing a general equilibrium model for the Russian economy in the transition process was to develop a quantitative analytical device based on sound theory, a consistent data set and which would be useful in explaining the development of Russia's agricultural sector in the transition process. Thereby, we also addressed the first research question which was posed in chapter 1: how useful can a general equilibrium framework be for the analysis? The expectation was that such a model would help to identify important factors that drive agricultural development in the given country context.

In a similar way to other country studies of this kind, this necessitated examination of the economic environment, the conditions in factor markets, and the impact of technologies which can be assessed in a general equilibrium framework (e.g. MUNDLAK 2000). As a result of the major structural break associated with the switch from plan to market, an econometric GE model as developed and used by Mundlak for country studies on Argentina and Chile would not have performed the task. Instead, we used a *computable general equilibrium* model that is comparative-static and which has been compiled for 1994.

Various *arguments* have been presented *in favor of using the computable general equilibrium* approach to examine the *effects of transition-related forces on Russia's agriculture*. In the course of transition markets, and hence prices, have become the central allocation mechanism in Russia's economy. The role of prices is likely to gain in importance as the institutional framework essential for markets to function efficiently is developed further. In the country context of Russia, one other important argument in favor of using the framework relates to the *complex linkages between agriculture and the macroeconomy*. These linkages are also sizable because Russia's agro-food sector is the second most important sector of the domestic economy: it employs a multitude of economic agents in production, processing, and retailing, and has superior importance with respect to consumer expenditures. The discussion of the results of simulations showed that a general equilibrium model can be quite useful in highlighting important economy-wide repercussions between sector developments and the macroeconomy and in replicating real world developments. Another argument relates to theoretical considerations: the *general equilibrium framework offers various alternatives to analyze different price effects*. The discussion in chapter 2 indicated that many institutional changes in the transition process, which are the result of the mix and the timing and sequencing of reforms in a given country context, can be analyzed by looking at the associated relative price changes. A third argument refers to the *great demand for quantitative policy analysis of economy-wide effects* in transition countries in general and in Russia in particular. Many strategic policy decisions, such as on the direction of future trade policies or the accession to the WTO, still have to be made. Against this background, policy-makers are demanding not only a quantitative assessment of sectoral effects but also of economy-wide effects.

The potential of CGE models was exploited further in this study by modifying the standard framework in such a way that it represents a structural model. Various *features of this structural model framework* were valuable in explaining the economy-wide responses to exogenous shocks. For instance, it was shown that demand factors were more important in influencing the way the economy responded to exogenous shocks. Sectors with high household expenditure shares, for example, responded particularly strongly to shocks which effectively yielded strong declines or increases in the income available in the economy. Due to the open labor market, adjustment to exogenous shocks took place not so much by means of far-reaching restructuring but instead by releasing labor when the economy contracted and vice versa.

Another promising option the modeling framework offers is the fact that it is based on a consistent theoretical and empirical framework. The *strict rules which governed the compilation of the data base* required the adaptation of data from secondary sources for some indicators in order to make it compatible with the theoretical constraints of social accounting. In view of the insecure data base, the use of this rigorous theoretical concept proved to be a valuable tool in modeling Russia's economy.

However, caution is called for in interpreting the results of these simulations. The model output must always be interpreted against the model inputs and the specific economic situation, as revealed by the benchmark solution which was specified for 1994 and which has some distinct features. For example, the high trade surplus Russia had in 1994 because of its substantial energy exports concealed the pressure on its currency that evolved in the mid–1990s. Or, the low level of exports in the agricultural sectors does not allow for tremendous export growth in our short run model, even if substantial export subsidies are granted. Furthermore, as mentioned above, we had to abstract from the *virtual economy*, which is characterized by payment arrears, barter trade, and debt defaults and/or soft budget constraints. While the economic relevance and the damage this *virtual economy* has caused has been widely acknowledged (GADDY and ICKES 1999), empirical information on its scope is too limited to incorporate it in the framework we have used in this study. Therefore, the model structure and data base were made as transparent as possible. A secondary effect is that it can now serve as a reference for further model developments.

Furthermore, we showed that the specific *closure rules* with which central features of the structural model were specified do indeed have significant effects on the results of simulations. Removing structural rigidities in the economy, which enables productive resources to move to their most efficient use in the economy, yields the most positive effects. At the same time, removing restrictions in factor markets, both capital and labor, should receive higher priority by the government in the reform period to come. In contrast, sensitivity analysis showed that in the structural modeling framework quite substantial alterations of trade elasticities would be mandatory to produce sizable changes in the economic results of respective experiments.

In spite of these modifications, skepticism arises regarding the use of the general equilibrium approach because of the insecure data base and the many obvious market imperfections in Russia's economy which were not explicitly taken into account.

For instance, the *behavioral differences* between small-scale and large-scale agricultural producers often relate to non-economic or institutional factors and certainly are *often too complex* to be represented in such an economy-wide framework.[1] Another limitation of the approach to CGE modeling chosen in this study refers to the fact that we carried out a *comparative static analysis*, which means that dynamic processes that might result in structural changes even within the transition process are not captured. In our model specification, we precluded more

---

[1]   For instance, MURPHY, VISHNY and SHLEIFER (1993) showed with a very simple two-sector model of a farm economy how rent-seeking can be very costly to growth. They argue that rent-seeking activities exhibit increasing returns which can lead to multiple equilibria. A 'bad equilibrium' with high rent-seeking activities may reduce productivity and protection of property rights. Based on the rationale of their model, MURPHY, VISHNY and SHLEIFER (1993: 412) argue that these effects of rent-seeking may actually describe what has happened during the collapse of communism in Russia.

flexible adjustments both on the production and on the demand side. Moreover, the *parameters* used in the model *are fixed* and not subject to change. Therefore, even though the model simulations have been classified as short-run solutions, caveats remain in this respect.

Furthermore, the possibilities to model the *complex micro-economic behavior of agents* in the transition process in such a framework remain limited. For instance, while some features typical of *Russia's labor markets*, such as increasing unemployment, have been captured with the model, other aspects of the country's labor markets were not. Even though the modeling framework provides options to improve the representation of labor markets as more data on sectoral characteristics become available, one of the most typical features of labor markets for the Russian Federation and other countries of the FSU can hardly be revealed within the framework: wage arrears have become a common phenomenon and result in behavior that might be different from that implicit in our model economy.[2] In conclusion, it is obvious that the CGE approach should not be considered as a tool that can address all aspects relevant to the economic policy analysis of transition economies but that it can be useful only in identifying important aspects of the transition process.

Possible extensions of the model

Having discussed the basic potentials and general limitations of the CGE framework, various extensions of the model we developed in this study could be considered in future work. In general, they relate to the following three dimensions:

Theoretical issues. To enhance the explanatory power of the model, different functional forms could be tested in future versions of the model. Another area providing scope for further work on theoretical issues is the representation of additional transition-specific features such as the soft budget constraints of large-scale farms, the implications of barter-arrangements (cf. HAAPARANTA and KERKELÄ 2000), or, more generally, the role of transaction costs. The latter could be addressed by introducing different marketing margins for domestic and foreign sales, which are also different between sectors. Furthermore, in chapter 5 we mentioned that the results of the experiments in the previous section were still not close enough to real world developments in so far as in the course of the first transition decade, an increase in output from small-scale agriculture was observed. By simply introducing a positive shift effect in this sector in one of the subsequent experiments, we were able to show with the model the increase in output in the small-scale sector in a situation in which all other sectors suffered (see Table 5-8). However, factors other than shifts in the supply curve, for instance, changing behavior on the demand side, might have driven the positive output response in the subsistence sector. Such features could be taken into account in future extensions of the model via various

---

[2] This view is supported by a statement made by STONEMAN (2000: 10): " ... the pervasiveness and institutionalization of non-payments is the single most important impediment to the emergence of an operable market economy [in Russia]."

alternative options. For instance, by explicitly differentiating between non-market and market transactions, which could be reflected by using the distinction between an activity account (production) and one commodity account (market) for each sector. Another option would be to allow for more flexibility in the behavior of consumers, and thereby achieving a higher rate of substitutability between the goods produced in the various sectors.

Data.   First, the data base can obviously be updated when consistent data sets, particularly input-output tables, for more recent base years become available; second, more detailed data on various components of the model such as households, labor and investment markets, regional disparities and the role of transport costs and marketing margins etc. could be incorporated; third, empirically estimated parameters for the transition period should be included in the data base whenever they become available; fourth, in addition to the representation of subsistence food production, more segments of the 'virtual' or 'hidden' economy could be taken into account more explicitly, as they represent an important share in the economy; fifth, the comparative static specification could be substituted for by a dynamic version to be able to better take account of long-term growth effects and the long-term potential of Russia's agro-food sector in world food markets.

Policy detail.   It is likely that macroeconomic distortions will be reduced further in the future course of transition and policies will be based on better defined legal terms and, hence, become more rational. Then, more detailed representation of specific policy instruments will become increasingly important. For instance, the representation of subsidized credit schemes as well as the division of labor between federal and regional government entities will allow better insight in the complex links between the country's macroeconomy and its agro-food sector.

   The wealth of potential areas in which there is scope for improvement of the model indicates that each modeling approach has to be selective. Even though the possibilities for modeling the Russian economy are likely to improve, it will remain important to gear the model's design to the specific research and policy questions.

Discussion of model results and policy conclusions

Agriculture and the food industry play a pivotal role in the Russian economy. The structural computable general equilibrium model we developed in chapters 2–4 was used in chapter 5 to simulate various economic events and policy changes and their effects on Russia's agro-food sector. While this model provided further insight into the repercussions between important economy-wide developments and the performance of the agro-food sector, it should not be expected that any such model can include all the effects of the multitude of reforms. If this were possible, *central planning* would have succeeded over the market paradigm and not the other way around. Some of the experiments were carried out using both the 1990 and the updated 1994 data base of the model, while others were carried out using only the 1994 version of the model. The model simulations were used to identify and discuss various economic mechanisms.

Results related to research question 2: explaining the poor economic performance of agriculture in the transition process and potential solutions

One set of simulations addressed the *second research question* which was raised in chapter 1: How can the weak economic performance of Russia's agro-food sector within the country and in comparison with other transition economies, particularly in central Europe, and, hence, the slowness of recovery from the original transition-specific shock be explained; and, what are potential ways out of this disastrous situation? This is also a question of concern for the GOVERNMENT OF THE RUSSIAN FEDERATION (2000: 6) which in mid–2000 launched a new agricultural policy in an attempt to stop the output decline in Russian agriculture in the immediate future. Many of the following conclusions directly relate to the areas in which the Government of Russia wishes to take action.

From our analysis, the following conclusions can be drawn with respect to this issue:

Agricultural output was strongly affected by deteriorating economy-wide conditions. Sector-neutral events can be far more important than often expected. The experiments showed the dependence of the agro-food sector on favorable economy-wide conditions. Investments, which would help to raise the efficiency of the transport sector, would be particularly important in order to reduce the costs associated with moving food efficiently through the domestic marketing system. In a country of Russia's size, this sector is particularly important because of the often vast distances between production and consumption locations.

Hence, the various experiments conducted highlighted that the severity of the output decline in Russia's large-scale agriculture can be explained best by the simultaneity of various economic and political processes. It would be mistaken to argue that single factors such as negative developments in agriculture's terms of trade were the only factors.

Agriculture's output decline contributed to the appreciation of the exchange rate. The high import share of food in Russia during the 1990s can be partially attributed to domestic factors. Various experiments, in which a deterioration of the capital stock or a declining productivity in the two large-scale agricultural sectors were simulated, showed that these factors directly reduced supply of the domestic good, which, under a flexible exchange rate system, effectively contributed to the appreciation of the ruble.

Agricultural sector policies which attempt to increase the total factor productivity in various segments of Russia's agricultural sector can have very different results. Based on our data base for 1994, we showed that increases in TFP achieved by large agricultural farms specializing in crop production yield higher positive output responses than those in the livestock sector. These results highlight that it is important to take the factor intensity in individual sectors into consideration when designing policies that intend to improve factor productivity. Linked with the increase in factor efficiency is another important observation which has been proven in many other countries: as TFP increases, additional supply is brought to the market. C.p., this will reduce the producer prices in the sector and set 'Cochrane's treadmill' into motion. Russian policy-makers should be aware that this process is the result of market forces and should not be used as an argument for increasing support to agricultural producers.

Positive output development in agriculture can be conducive to overall economic growth. In fact, the same experiment clearly showed two economy-wide effects that MUNDLAK (2000) referred to as the two major contributions of agriculture to economic development: the increase in supply reduces not only producer prices. The positive side of the same coin is *the decrease in consumer prices for food,* which contributes to significant welfare gains on the consumer side because of the high food expenditure shares of the average Russian household. Furthermore, an increase in productivity in agriculture releases labor from this sector which can then be employed in other sectors.

Sectors with high household expenditure shares are hurt most by a reduction in income in the economy, but they may also benefit more than proportionally from an increase in output. Due to the very high share of average households' expenditures being spent on food, the growth potential of Russian food industries and agriculture is significant. The same holds true for the light manufacturing sector, which has the highest share of household expenditures and which is therefore likely to benefit from economic growth as well.

Small-scale agriculture was significant for output stability and domestic food security in the transition period. Interesting results were highlighted with respect to the small-scale agricultural producers. For instance, while the relatively low intensity of the use of industrial intermediates in the production process of small-scale agricultural producers insulates the small-scale sectors to some extent, its important role with respect to household expenditures keeps these sectors exposed to economy-wide developments. Therefore, the role of the LPH sector and, hence, the small-scale agricultural producers for food security should not be underestimated. For example, the simulations with the 1994 model version highlighted that small-scale agriculture was in fact a buffer against negative terms of trade shocks. Having been widely insulated from price developments on world markets, it was this sector that kept production not only more or less constant – as shown in the simulations – but even increased output. A similar mechanism was observed for other negative economy-wide effects, for example the deterioration of the transport sector and – from the point of view of large-scale agriculture – for negative sector policies such as a reduction in government policies. We showed that both policies had rather little effect on small-scale agriculture, which was effectively sheltered from such events because of its subsistence character.

Commercialization of the most efficient small-scale agricultural producers should become part of Russia's agricultural policy. While small-scale agriculture might have been a buffer against the adverse effects of the initial reform period, its role in the years to come should be reconsidered. Particularly the degree of commercialization of the different farm types in Russia will be crucial for the future development of the whole sector. One simulation indicated that differences in the degree of commercialization – modeled here by different degrees of forward and backward linkages – have important implications. It was shown that the higher the degree of commercialization of a specific segment of agricultural production, the more this sector will benefit from economy-wide growth. Hence, the degree of commercialization will be an important determinant for the degree to which the various farm types will be able to benefit from improved economy-wide conditions. If the opportunities for the small-scale farms to reach the market are not improved, a significant portion of Russia's agricultural production will continue to be driven by subsistence production and not by market factors. Hence, the positive relationship between commercialization of the small-scale farm sector and economic developments in countries in which this sector produces major portions of agricultural output should not be neglected (cf. VON BRAUN, DE HAEN and BLANKEN 1991). This is even more important with respect to the wide diversity of the agro-food sector in the various regions of Russia (KUHN and WEHRHEIM 1999). While there are many regions within Russia that are characterized by a high share of large-scale farms, there are other regions in which subsistence production is much more prevalent. Unless the options for commercialization of efficient and successful small-scale farmers in these regions are enhanced, the economic situation in the rural economies of Russia's regions is likely to diverge further. In fact, the new agricultural policy pursed by the Russian government acknowledges the importance of commercializing efficient small-scale agricultural producers in an attempt to improve the living conditions of the rural populace. It recommended that local authorities, consumer co-operatives, and large-scale agricultural enterprises should support private small-scale farms in the production process, in marketing their output, and in establishing cooperatives to provide them with services (GOVERNMENT OF RUSSIA 2000: 14).

Results related to research question 3: Russia's foreign trade strategies and the agro-food sector

A second group of simulations highlighted the effects of sector-neutral shocks to which the whole Russian economy was exposed as a result of the newly gained openness of its trade regime. Thereby these simulations referred to the *third research question* which was also phrased in chapter 1: what are the effects of various foreign trade strategies of the Russian Federation and how will they affect the country's agro-food sectors?

Negative terms of trade shocks which yielded a strong depreciation of the exchange rate did not result in overall economic growth. It was shown that the short run effects of a depreciation of the exchange rate were negative for the agro-food sectors as an aggregate, as well as for the economy as a whole. Various factors were identified as having driven these results: the high import dependence of agro-food sectors in the base period; a limited possibility to restructure within the economy towards the export-oriented sectors; the significant extent of capital flight in the base period, and the limited potential to commercialize important segments of Russia's primary agriculture. These findings complement the major conclusion in a paper on

Russia's financial crisis (SEROVA, VON BRAUN and WEHRHEIM 1999): a strong real devaluation of the ruble without structural reforms in Russia's agro-food sector will not turn around the poor performance of Russia's agro-food sector. In one experiment, the closure rule was changed is such a way that investments were flexible and foreign savings could adjust to terms of trade shocks. The experiment showed that additional income which resulted, for example from additional export revenues of the raw material sectors, is 'exported' and is not reinvested. If such behavior were to determine the domestic investment climate, the agro-food sectors would continue to suffer from capital shortages, a situation that has been prevalent for long in the 1990s.

The restructuring of Russia's agro-food trade in the transition period was not only caused by the liberalization of Russia's trade regime but also to a significant extent by the appreciation of the exchange rate. The model simulations showed that declining productivity and the deteriorating capital stock in large-scale agriculture were one factor that contributed to the appreciation of the real exchange rate. Hence, while the primary effect of declining productivity and declining capital stock in large-scale agriculture was to reduce domestic supply, these sector-specific developments also affected macroeconomic variables. In fact, the appreciation of the exchange rate effects made imports of food products cheaper and exports of agricultural commodities more expensive and, thereby, induced further restructuring of Russia's agro-food sectors towards imported food commodities. Additionally, the real appreciation has contributed to the depression in Russia's industrial heartland in the mid–1990s (BERGLÖFT and VAITILIGNAM 1999). Obviously other factors contributed to the real appreciation of the ruble between 1993 and 1997. Macroeconomic stabilization policies were certainly one driving factor causing the real appreciation of the ruble. In fact, the detrimental effects of an appreciation of the ruble as they were shown with our model and as they were experienced by Russia in the period preceding the financial crisis indicate that macroeconomic stabilization can come at high costs if macroeconomic balances (e.g. with respect to the exchange rate) are not maintained.

WTO-accession will be important to prevent detrimental effects from increasing preferential treatment of agriculture. If Russia does not join the WTO, demand for protection of agriculture is likely to increase. However, the simulations document that support granted to the farm sectors and the food industries comes at a price which the other sectors of the economy – the households, or the taxpayers – have to pay.

If support to agriculture is increased, direct domestic support should be favored. Russia's current position in the accession negotiations is to maintain the options to grant its agro-food sector protection with three policy components: domestic support, import tariffs, and export subsidies. Increasing the import tariffs in all agro-food sectors or introducing export subsidies can – under a flexible exchange rate system – induce indirect effects because of alterations in the exchange rate. In contrast, support to agriculture using direct domestic support measures does not have such significant distortionary effects on the exchange rate. In our simulations such policies, which are classified in the WTO context as 'blue box measures', revealed the least serious economy-wide repercussions.

It should also be born in mind that any export subsidy is effectively associated with a transfer of an economic rent from domestic taxpayers to foreign consumers. One of the simulations on the discussion of the various policy instruments available for supporting agriculture showed that the large-scale agricultural sector would significantly increase exports, at least in relative terms. If agricultural exports from Russia were to be more substantial in the future and then be supported with export subsidies, such a policy might be associated with substantial payments from the government. Effectively, this would mean that Russian taxpayers contribute to lower prices of the exported agricultural commodities consumed elsewhere in the world.

Liberalization of the agro-food sector may not yield the expected positive benefit as long as the economy is not fully flexible to shift the respective resources to the relatively more competitive sectors. One simulation showed a seemingly counter-intuitive result: GDP declined when the agro-food sectors were fully liberalized. Within our model economy, this effect can be explained by the fact that one production factor, namely capital, was not allowed to move inter-sectorally, which meant that the non-food sectors, whose competitiveness increased, could not grow sufficiently in order to compensate for the output decline in the agro-food sectors. In order to exploit the full potential of liberalization, it will therefore be essential to ensure that capital markets are as flexible as possible and that efficient rural financial markets will emerge. This causality is likely to be valid with respect to all production factors. In fact, as NORTH argues, it is the flexibility of economic institutions in market economies that has enabled them to achieve the productivity improvements of the past centuries. And, ironically, it has been the inflexibility and rigidities of centrally planned economies that have led to their demise (NORTH 1995: 22).

However, even against the background of the inflexible economy, it was shown that the higher the degree of overall liberalization, the more the Russian economy as a whole will benefit. Another simulation based on the structural version of Russia's model economy examined the effects of a complete abolition of Russia's import tariffs. Here, relatively high positive effects on GDP and on employment were the result. These effects are only possible because of the disequilibrium notion of the model: as a result of the increased competitiveness of domestic production sectors, liberalization not only results in a restructuring of domestic production but also coincides with an expansionary effect that can be achieved because more labor is attracted into the official economy.

The introduction of a uniform import tariff rate would be most beneficial to Russia's economy if it coincided with an effective overall cut in import tariffs. Furthermore, we simulated the replacement of varying import tariffs with a uniform import tariff or flat rate. The economic effects that can be revealed with our model economy quite obviously depend on the level of the uniform tariff rate. The lower the tariff rate, the greater the benefits for the whole economy. These immediate economic effects of liberalizing the trade regime are likely to be complemented by economic gains because of reduced opportunities for corruption and the associated losses in economic efficiency.

Summing up, the study highlights that *modeling Russia's economy in transition* can contribute significantly to a better understanding of the economic repercussions between various sectors. Macroeconomic developments strongly influenced the evolution of the Russian agro-food sectors in the 1990s and *vice versa*. Therefore, the applied general equilibrium model that has been described in this study is a first step in developing a modeling framework for Russia's economy with which a wide range of policy analyses can be facilitated. In fact, we showed that the use of an applied general equilibrium model can serve various purposes in the context of transition economies: first, the economy-wide model can be a useful tool for data management in spite or because of the surmounting data problems in respective countries and thereby can serve as a policy information systems by itself. Second, and more importantly, it can also be helpful in analyzing and understanding some of the economic peculiarities of the transition process or counterintuitive effects of policies in the context of a transition economy. Furthermore, it was shown that basic economic thought has much to offer in explaining the transition path of an individual country like Russia. As it is likely that the transition process will continue for many years to come, the rigorous application and modification of available economic research tools for quantitative policy analysis will continue to be of pivotal importance.

Appendix

Table A2-1 The equations of the 1–2–3 model

Flows	Prices
(1) $\overline{X} = G(E, D^S; \Omega)$	(7) $P^m = R \cdot pw^m$
(2) $Q^S = F(M, D^D; \sigma)$	(8) $P^e = R \cdot pw^e$
(3) $Q^D = \dfrac{Y}{P^q}$	(9) $P^x = g_1(P^e, P^d)$
(4) $\dfrac{E}{D^S} = g_2(P^e, P^d)$	(10) $P^q = f_1(P^m, P^d)$
(5) $\dfrac{M}{D^S} = f_2(P^m, P^d)$	(11) $R \equiv 1$
(6) $Y = P^x \cdot \overline{X} + R \cdot \overline{B}$	
Equilibrium conditions	**Identities**
(12) $D^D - D^S = 0$	(i) $P^x \cdot \overline{X} \equiv P^e \cdot E + P^d \cdot D^S$
(13) $Q^D - Q^S = 0$	(ii) $P^q \cdot Q^S \equiv P^m \cdot M + P^d \cdot D^D$
(14) $pw^m \cdot M - pw^e \cdot E = \overline{B}$	(iii) $Y \equiv P^q \cdot Q^D$
Endogenous Variables	
E: Export good	P^x: Price of aggregate output
M: Import good	P^q: Price of composite good
D^S: Supply of domestic good	R: Exchange Rate
D^D: Demand for domestic good	**Exogenous Variables**
Q^S: Supply of composite good	pw^e: World price of export good
Q^D: Demand for composite good	pw^m: World price of import good
Y: Total income	B : Balance of trade
P^e: Domestic price of export good	X : Aggregate output
P^m: Domestic price of import good	σ: Import substitution elasticity
P^d: Domestic price of domestic good	Ω: Export transformation elasticity

Source: DE MELO and ROBINSON (1989).

Table A3-1   The complete set of equations of the Russia model

No.	Name of equation	Equation	No. of equations in the model
		Production, and demand for primary factors and intermediates	
	Demand for intermediates	$V_i = \Sigma_j a_{ij} X_j$	n
	Production	$X_i = a_i^x \Pi_f F_{if}^{\alpha_{if}}$	n
	Factor demand	$W_f \phi_{if} := \cdot P_i^v \cdot \alpha_{if} \cdot \dfrac{X_i}{F_{if}}$	$f \cdot n$
	Commodity supply	$X_i = a_i [\delta_i E^{\rho_i} + (1 - \delta_i) D_i^{\rho_i}]^{1/\rho_i}$	n
	Producer price	$P_i^x = \dfrac{P_i^d \cdot D_i + P_i^e \cdot E_i}{X_i}$	n
	Net price	$P_j^v = P_j^x (1 - t_j^x) - \Sigma_i a_{ij} P_i^q$	n
	Demand for investment goods	$I_i^A = \Sigma_j b_{ij} \cdot I_j^B$	n
	Capital good price	$P_j^k = \Sigma_i b_{ij} P_i^q \ with \Sigma_i b_{ij} = 1$	n
	Change in warehouse stocks	$I_i^W = \beta_i^t X_i$	n
	Gross nominal investments	$\tilde{I}^B = \tilde{I} - \Sigma_i \beta_i^I X_i$	1
	Real investments in sector j	$I_j^B = \beta_j^B \tilde{I}^B / P_j^k \ with \Sigma_j \beta_j^B = 1$	n
		Allocation of income	
	Factor income	$\tilde{Y}_f = \Sigma_i W_f \cdot \phi_{if} \cdot F_{if}$	f
	Net income of capital owners	$\tilde{Y}^K = \tilde{Y}_{f=1} - \tilde{A} - \tilde{T}_{f=1}^f$	1
	Depreciation	$\tilde{A} = \Sigma_i d_i \cdot P_i^k \cdot F_{i1}$	1
	Net income from labor	$\tilde{Y}^L = \Sigma_{f \neq 1} \tilde{Y}_f$	1
		Household behavior	
	Private consumption	$P_i^q \cdot C_i = \Sigma_p \gamma^{P_i} \cdot \tilde{Y}_p \cdot (1 - t_p) \cdot (1 - s_p)$	n
	Consumer price	$P_i^q = \dfrac{P_i^d \cdot D_i + P_i^m \cdot M_i}{Q_i}$	n

Household savings	$\widetilde{S}_p = \Sigma_p s_p \cdot (1-t_p) \cdot \widetilde{Y}_p$	p

Transactions with the rest of the world

Export supply	$E_i^S = D_i [\dfrac{(1-\delta_i^t)}{\delta_i^t} \dfrac{P_i^e}{P_i^d}]^{\sigma_i^t} \ with \sigma_i^t = \dfrac{1}{\rho_i^t - 1}$	n
Price of exports	$P_i^e = P_i^{\$e}(1+t_i^e)R$	n
Export demand	$E_i^D = \overline{E}_i (P_i^{\$e} / \overline{P}_i^{\$e})^{-\eta_i^t}$	n
Price of imports	$P_i^m = \overline{P}_i^m (1+t_i^m)R$	n
Import demand	$M_i = D_i [\dfrac{\delta_i^q)}{(1-\delta_i^q)} \dfrac{P_i^d}{P_i^m}]^{\sigma_i^q} \ with \sigma_i^q = \dfrac{1}{1+\rho_i^q}$	n
Commodity demand	$Q_i = a_i^q [\delta_i^q M_i^{-\rho_i^q} + (1-\delta_i^q)D_i^{-\rho_i^q}]^{-1/\rho_i^q}$	n

Government behavior

Custom tariff income	$\widetilde{T}^m = \Sigma_i t_i^m \cdot P_i^m \cdot R \cdot M_i$	1
Revenues from export taxes	$\widetilde{T}^e = \Sigma_i t_i^e \cdot P_i^e \cdot R \cdot E_i$	1
Revenues from consumption taxes	$\widetilde{T}^x = \Sigma_i t_i^x \cdot P_i^x \cdot X_i$	1
Total government revenues	$\widetilde{Y}^G = \widetilde{T}^m + \widetilde{T}^e + \widetilde{T}^x$	1
Government consumption	$C_i^G = \gamma_i^G \cdot \overline{C}^G$	n
Government savings	$\widetilde{S}^G = \widetilde{Y}^G - \Sigma_i P_i^q \cdot C_i^G$	1

Savings

Total savings	$\widetilde{S} = \widetilde{S}_p + \widetilde{S}^G + \widetilde{A} + \widetilde{S}^F$	1

Gross domestic product

Nominal GDP	$GDP^n = \Sigma_i P_i^v \cdot X_i + \widetilde{T}^x + \widetilde{T}^m - \widetilde{T}^e$	1
Real GDP	$GDP^r = \Sigma_i (C_i + C_i^G + I_i^A + I_i^W + E_i - \overline{P}_i^{\$m} \cdot M_i \cdot R)$	1
Price index	$\overline{P} = GDP^n / GDP^r$	1

Closure of the model

Comodity market equilibrium	$Q_1 = V_i + C_{ip} + C_i^G + I_i^A + I_i^W$	n
Factor market equilibrium	$\overline{F}_f = \Sigma_i F_{if}$	f
Trade balance	$\Sigma_i \overline{P}^{\$m} \cdot M_i = \Sigma_i P^{\$e} \cdot E_i + \overline{S}^F$	1
Savings-investment identity	$\widetilde{S} = \widetilde{I}$	1

	Number of equations
	$= 19 \cdot n + f \cdot n + 2\,f + p + 15 \cdot$
with n=20, f=2, and p=1	$= 19 \cdot 20 + 2 \cdot 20 + 2\,2 + 1 + 15 \cdot \ = 440$

Table A3-2 The variables in the Russia model

Variable Endogenous	Exogenous	No.	Name of variable
P_i^m		n	Consumer price of imported commodity
	$\bar{P}_i^{\$m}$	n	Cif-import price of imported commodity in foreign currency
R		one	Exchange rate
P_i^e		n	Producer price of exported good
$P_i^{\$e}$		n	Fob-export price of exported good in foreign currency
P_i^q		n	Consumer price of composite good
P_i^d		n	Price of commodity produced and consumed domestically
D_i		n	Quantity supplied and demanded of commodity produced and consumed domestically
M_i		n	Quantity of the imported good
Q_i		n	Quantity demanded of the composite consumption good (consisting of domestic and imported good)
P_i^x		n	Producer price of the composite good (domestic sales and exports)
E_i		n	Supplied and demanded quantity of export commodity
X_i		n	Production volume in sector
P_j^v		n	Price of value added of commodity
P_j^k		n	Price of the composite capital good in sector
	$\bar{P}$	one	Deflator of Gross Domestic Product
F_{fi}		$f \cdot n$	Demand for factor f in sector

W_f		f	Average wage or interest rate for factor f
V_i		N	Total intermediate demand for composite good
	$\bar{P}_i^{\$e}$	N	Average world market price of export commodity
	$\bar{E}_i$	n	Exogenous export demand for good
C_i		n	Private demand for composite good
C_i^G		n	Real government consumption of good
	$\bar{C}^G$	one	Real government consumption
I_i^w	0	n	Change in warehouse stocks
$\tilde{I}^B$		one	Total nominal gross investments
$\tilde{I}$		one	Total nominal investments (including change in warehouse stocks)
I_j^B		n	Real gross investments in sector
I_i^A		n	Total demand for investment good

	$\bar{F}_f$	f	Factor supply
	$\bar{S}^F$	one	Foreign savings in foreign currency
$\tilde{S}_p$		p	Savings of households
$\tilde{S}$		one	Total savings (domestic and foreign savings)
$\tilde{A}$		one	Depreciation
$\tilde{Y}_f$		f	Factor income of factor f
$\tilde{Y}_p^K$		one	Income of capital owners
$\tilde{Y}_p^A$		one	Income of workers
$\tilde{T}^m$		one	Total customs tariff revenue of the government
$\tilde{T}^e$		one	Total expenditures of the government for export subsidies
$\tilde{T}^x$		one	Total sales or consumption tax revenues of the government

$\tilde{Y}^G$	one	Total government revenues minus export subsidies
$\tilde{S}^G$	one	Government budget deficit
GDP^n	one	Nominal gross domestic product
GDP^r	one	Real gross domestic product
No. of endogenous variables with n=20, f=2, and p=1		$= 19 \cdot n + f \cdot n + 2f + p + 14 \cdot$ $= 19 \cdot 20 + 2 \cdot 20 + 22 + 1 + 17 \cdot = 439$

Table A3-3   Overview on notation in the model

Abbreviation examples	Type	Explanation for abbreviation
$\bar{P}$	Capital letter with a bar	Variable is exogenous
X, C	Capital letter	Endogenous variable; e.g. volume of domestic production and consumption, respectively
d	Greek letter	Shares of respective variable in base run situation
a, t	Small letter	Policy variable or parameter
$\tilde{S}$	Capital letter	Variable is nominal
m, e, d	Superscript	Indicating the origin or destination of the respective product/price: imported, exported, and domestic
$\$$	Superscript	Value given in foreign currency
i and j	Subscript	Sector in the economy; i denotes the originating and j the receiving sector
F	Subscript	Production factors (labor or capital)
P	Subscript	Household variable
x, v, h, f	Superscript	Indicating different types of tax revenues (consumption, value added, household income and factor taxes)
G	Superscript	Government
W, B, A	Superscript	Type of investments (warehouse stocks, gross investments, depreciation)
n, r	Superscripts	Nominal or real variable

Table A3-4 **Additional abbreviations used for exogenous shares and parameters**

Variables	Name of variable
	Shares
δ_i^t	Split of the production in sector i in the reference period between production for domestic and export market
γ_i^p, γ_i^G	Household and government consumption shares
b_{ij}	Capital good invested in the base run
β_i^t	Stocks in the base run
	Elasticities and efficiency parameters
α_{if}	Partial production elasticities of production factors
	Elasticity of transformation/substitution
η_{it}	Export demand elasticy
a_i^x	Total Factor Productivity parameter
	Rates
d	Depreciation rate
s_p	Households' savings rate or marginal propensity to save

Table A4-1   Stylized macro Social Accounting Matrix

	1 Activities	2 Commodities	3 Factors of production — Labor	3 Factors of production — Capital	4 Households	5 Government	6 Capital account	7 Rest of World	Sum
	1 ... n sectors	1 ... n sectors	1 ... z types of labor	Capital					
1 1 ... n sectors		Domestic sales				Export subsidies		Exports	*Total domestic revenues*
2 Commodities 1 ... n sectors	Intermediate demand				Household consumption	Government consumption	Investments		*Total domestic*
3 Factors of production Labor 1 ... z types of labour	Labor demand								*Labor income*
Capital	Capital demand								*Capital income*
4 Households			Labor enumeration	Capital costs					*Household income*
5 Government	Direct taxes	Import tariffs	Labor taxes	Direct taxes	Indirect Household taxes				*Government revenue*
7 Rest of World		Imports							*Revenues*
Sum	*Total domestic output*	*Total domestic demand*	*Labor costs*	*Capital payments*	*Household expenditures*	*Government expenditures*	*Investment expenditures*	*Exports*	

E X P E N D I T U R E S

R E V E N U E S

Table A4-2 Example of a stylized micro Social Accounting Matrix

Variable	Variable	*Expenditures of economic sectors* 1	..	i	**Total Outlays** TO	*Household demand* H	*Government demand* G	*Investment demand* I	**Domestic absorption** TA=TO+H+G+I	*Imports incl. tariffs* M	**Demand for domestic production** DD=TA-M	*Exports incl. taxes* E	**Use of domestic production** UDP= DD+E	**Balance** UDP-ODP=
Revenues of economic sectors	1				$\sum_j$				$\sum_i$		$\sum_i$		$\sum_i$	$\sum_i - \sum_j = 0$
	..				$\sum_j$				$\sum_i$		$\sum_i$		$\sum_i$	$\sum_i - \sum_j = 0$
	j				$\sum_j$				$\sum_i$		$\sum_i$		$\sum_i$	$\sum_i - \sum_j = 0$
Total expend. Intermediates	TEI	$\sum_j$	$\sum_j$	$\sum_j$	$\sum_i\sum_i$	$\sum_i$	$\sum_j$	$\sum_j$	$\sum_i\sum_j$	$\sum_j$	$\sum_i\sum_j$	$\sum_j$	$\sum_i\sum_i$	$\sum_i\sum_j - \sum_i\sum_i = 0$
Production factors														
Capital	C				$\sum_i$									
Labor	L				$\sum_i$									
Indirect taxes	IT				$\sum_i$									
Total value added	TVA=C +L+IT	$\sum_j$	$\sum_j$	$\sum_j$	$\sum_i\sum_i$									
Origin of domestic prod.	ODP=T EI+TVA	$\sum_j$	$\sum_j$	$\sum_j$	$\sum_i\sum_i$									

Table A4-3 Aggregation scheme for 1990 IOT of Russia

Aggregated sector	Numbers of Sectors out of 125-IOT for 1990
1. Electric power	1 Power
2. Fuel industry	2 Oil products, 3 Refineries, 4 Gas & gas products, 5 Coal
3. Metallurgy	6 Combustible shales, 7 Peat, 8 Ferrous ores, 9 Ferrous metals, 10 Coking products, 11 Fire resistant materials, 12 Metal products, 13 Non-ferrous ores, 14 Non-ferrous metals
4. Chemical industry	15 Mineral chemistry, 16 Basis chemicals, 17 Chemical fibers, 18 Synthetic resins, 19 Plastic products, 20 Paints & lacquers, 21 Synthetic paints, 22 Synthetic rubber, 23 Organic chemicals, 24 Tires, 25 rubber & asbestos, 26 Other chemical products
5. Machine building	27 Energy & power equipment, 28 Hoisting technology, 29 Mining, 30 Transportation, 31 Railway equipment, 32 Electrotechnical, 33 Cable products, 34 Pumps & chemical equipment, 35 Machine tools, 36 Forging/Pressing, 37 Casting, 38 Precision instruments, 39 Synthetic diamonds, 40 Tools and dies, 41 Autos & parts, 42 Bearings, 43 Tractors & agricultural, 44 Construction, 45 Communal, 46 Light industry, 47 Processed food, 48 Trade & dining, 49 Printing, 50 Household appliances, 51 Sanitary engineering, 52 Shipbuilding, 53 Radio electronics, 54 Other industries, 55 Metal construction, 56 Metal products, 57 Repair
6. Wood industry	58 Logging 59 Sawmills & Lumber, 60 Plywood, 61 Furniture, 62 Paper & Pulp, 63 Wood chemistry prod., 106 Agricultural services
7. Light manufacturing	72 Glass, 73 Cotton prod., 74 Flax prod., 75 Wool prod., 76 Silk prod., 77 Knitwear, 78 Other textiles, 79 Sewn goods, 80 Leather, 97 Pharmaceuticals, 98 Medical equipment, 99 Medical prod., 100 Other prod.
8. Construction	64 Cement, 65 Asbestos products, 66 Roofing & insulation, 67 Prefab concrete, 68 Wall materials, 69 Construction ceramics, 70 Linoleum products, 71 Other constr. materials, 102 Construction
9. Sugar	81 Sugar
10. Flour & bread	82 Bread and baked products, 95 Flour & cereals
11. Meat products	91 Meat products
12. Dairy products	92 Dairy products
13. Other food	83 Confections, 84 Vegetable oils, 85 Perfume oils, 86 Distilleries, 87 Wines, 88 Fruit & vegetables, 89 Tobacco, 90 Other food
14. Animal feed	96 Animal Feed
15. Agric. crops	104 Agricultural crops
16. Animal prod.	105 Animal husbandry
16. Trade & transport	108 Transport-cost, 109 Road services, 110 Communications, 111 Information Process., 112 Other prod. sectors,
17. Services	122 Housing-communal, 123 Non-productive transport, 124 Non-productive communic., 125 Education, 126 Culture, 127 Health & recreation, 129 Science, 130 Banking & insurance, 131 State admin., 133 Defense & other

Source: IOT for 1990 by WORLD BANK.

Table A4-4 Input Output Coefficients (for 1990) used in the CGE model for the Russian Federation

Sector	1	2	3	4	5	6	7	8	9	10	11	12	13	14	15	16	17
1	0.010	0.038	0.038	0.074	0.019	0.021	0.013	0.014	0.003	0.005	0.002	0.004	0.009	0.004	0.007	0.024	0.016
2	0.245	0.257	0.052	0.024	0.005	0.014	0.003	0.013	0.008	0.003	0.001	0.002	0.010	0.002	0.010	0.028	0.015
3	0.012	0.003	0.428	0.039	0.105	0.010	0.013	0.041	0.002	0.000	0.001	0.001	0.004	0.000	0.001	0.002	0.010
4	0.003	0.010	0.011	0.290	0.030	0.028	0.037	0.013	0.002	0.001	0.001	0.001	0.010	0.005	0.024	0.006	0.019
5	0.034	0.012	0.022	0.027	0.289	0.029	0.009	0.065	0.002	0.003	0.001	0.002	0.013	0.003	0.039	0.019	0.084
6	0.001	0.005	0.004	0.024	0.011	0.261	0.012	0.031	0.002	0.003	0.002	0.003	0.018	0.002	0.005	0.014	0.005
7	0.013	0.004	0.005	0.045	0.015	0.034	0.443	0.026	0.015	0.005	0.002	0.003	0.016	0.026	0.008	0.032	0.002
8	0.001	0.001	0.002	0.003	0.003	0.004	0.002	0.132	0.002	0.001	0.000	0.000	0.001	0.001	0.003	0.005	0.002
9	0.000	0.000	0.000	0.000	0.000	0.000	0.000	0.000	0.108	0.001	0.000	0.002	0.017	0.001	0.000	0.000	0.002
10	0.000	0.000	0.000	0.000	0.000	0.000	0.001	0.000	0.000	0.222	0.003	0.000	0.015	0.062	0.025	0.001	0.006
11	0.000	0.000	0.000	0.000	0.000	0.000	0.003	0.000	0.000	0.002	0.101	0.001	0.010	0.019	0.002	0.001	0.007
12	0.001	0.000	0.001	0.003	0.001	0.001	0.001	0.001	0.000	0.002	0.001	0.075	0.007	0.015	0.004	0.002	0.005
13	0.000	0.000	0.000	0.013	0.000	0.001	0.004	0.000	0.002	0.017	0.008	0.002	0.225	0.158	0.006	0.000	0.008
14	0.000	0.000	0.000	0.000	0.000	0.000	0.001	0.000	0.000	0.006	0.000	0.000	0.002	0.060	0.052	0.000	0.000
15	0.000	0.000	0.000	0.000	0.001	0.001	0.069	0.000	0.497	0.484	0.678	0.616	0.158	0.491	0.236	0.005	0.005
16	0.153	0.076	0.069	0.054	0.035	0.070	0.034	0.092	0.041	0.047	0.026	0.222	0.066	0.020	0.026	0.032	0.030
17	0.018	0.029	0.016	0.014	0.030	0.011	0.009	0.033	0.003	0.111	0.003	0.004	0.015	0.001	0.015	0.041	0.028
Sum	0.493	0.436	0.647	0.609	0.543	0.485	0.652	0.461	0.687	0.915	0.829	0.938	0.595	0.868	0.465	0.213	0.248

Notes: Names of Sectors: 1 Electric power; 2 Fuel industry; 3 Metal industry; 4 Chemical industry; 5 Machine building; 6 Wood industry; 7 Light manufacturing; 8 Construction; 9 Sugar refinery; 10 Flour milling; 11 Meat processing; 12 Dairy processing; 13 Other food; 14 Animal feed; 15 Agriculture; 19 Trade & Transport; 20 Services.

Source: Own calculations derived from IOT as published in WORLD BANK 1995.

Table A4-5 Input Output Coefficients for 1994 derived from up-dated macro-SAM (see Table A4-2)

Sec.	1	2	3	4	5	6	7	8	9	10	11	12	13	14	15	16	17	18	19	20
1	0.009	0.023	0.045	0.110	0.033	0.030	0.018	0.016	0.006	0.005	0.003	0.004	0.014	0.010	0.013	0.022	0.012	0.035	0.023	0.023
2	0.238	0.199	0.009	0.006	0.001	0.003	0.001	0.002	0.002	0.001	0.000	0.000	0.002	0.001	0.005	0.003	0.002	0.018	0.003	0.003
3	0.012	0.002	0.401	0.006	0.017	0.001	0.002	0.004	0.000	0.000	0.000	0.000	0.001	0.000	0.000	0.000	0.000	0.000	0.000	0.001
4	0.003	0.008	0.013	0.315	0.014	0.011	0.014	0.004	0.001	0.000	0.000	0.000	0.004	0.003	0.047	0.002	0.008	0.032	0.002	0.006
5	0.034	0.009	0.074	0.030	0.376	0.002	0.001	0.004	0.000	0.000	0.000	0.000	0.001	0.000	0.024	0.018	0.007	0.036	0.001	0.003
6	0.001	0.004	0.005	0.026	0.015	0.330	0.011	0.021	0.002	0.002	0.001	0.002	0.016	0.003	0.004	0.011	0.005	0.009	0.008	0.004
7	0.013	0.003	0.006	0.049	0.021	0.043	0.501	0.018	0.016	0.003	0.002	0.002	0.015	0.013	0.008	0.017	0.009	0.045	0.018	0.020
8	0.001	0.001	0.002	0.003	0.004	0.005	0.003	0.097	0.007	0.001	0.000	0.000	0.002	0.003	0.008	0.018	0.009	0.106	0.008	0.006
9	0.000	0.000	0.000	0.000	0.000	0.000	0.000	0.000	0.103	0.004	0.000	0.002	0.021	0.001	0.000	0.001	0.000	0.000	0.000	0.002
10	0.000	0.000	0.000	0.000	0.000	0.000	0.001	0.000	0.000	0.176	0.002	0.000	0.018	0.109	0.000	0.036	0.000	0.000	0.000	0.011
11	0.000	0.000	0.000	0.000	0.000	0.000	0.003	0.000	0.000	0.002	0.100	0.000	0.011	0.030	0.000	0.007	0.000	0.000	0.001	0.007
12	0.001	0.000	0.001	0.003	0.001	0.001	0.001	0.001	0.000	0.002	0.001	0.059	0.008	0.025	0.000	0.013	0.000	0.000	0.001	0.006
13	0.000	0.000	0.000	0.014	0.001	0.001	0.004	0.001	0.001	0.013	0.008	0.002	0.264	0.065	0.000	0.005	0.000	0.034	0.000	0.002
14	0.000	0.000	0.000	0.000	0.000	0.000	0.001	0.000	0.000	0.005	0.000	0.000	0.002	0.196	0.000	0.092	0.026	0.000	0.000	0.000
15	0.000	0.000	0.000	0.000	0.000	0.000	0.014	0.000	0.453	0.615	0.008	0.000	0.104	0.209	0.051	0.024	0.010	0.000	0.000	0.003
16	0.000	0.000	0.000	0.000	0.000	0.000	0.009	0.000	0.000	0.002	0.300	0.296	0.003	0.113	0.039	0.139	0.011	0.006	0.000	0.003
17	0.000	0.000	0.000	0.000	0.000	0.000	0.004	0.000	0.025	0.007	0.013	0.096	0.028	0.007	0.068	0.096	0.117	0.000	0.002	0.000
18	0.003	0.001	0.000	0.000	0.000	0.000	0.001	0.000	0.040	0.012	0.055	0.012	0.003	0.011	0.003	0.005	0.000	0.069	0.002	0.002
19	0.149	0.059	0.132	0.059	0.069	0.088	0.063	0.132	0.040	0.037	0.230	0.101	0.097	0.058	0.114	0.102	0.014	0.007	0.122	0.043
20	0.018	0.022	0.051	0.015	0.063	0.014	0.035	0.138	0.005	0.009	0.122	0.003	0.055	0.048	0.089	0.038	0.003	0.002	0.049	0.046
Sum	0.480	0.332	0.740	0.637	0.619	0.531	0.686	0.437	0.701	0.895	0.835	0.584	0.669	0.905	0.473	0.653	0.236	0.397	0.239	0.192

Notes: Names of Sectors: 1 Electric power; 2 Fuel industry; 3 Metal industry; 4 Chemical industry; 5 Machine building; 6 Wood industry; 7 Light manufacturing; 8 Construction; 9 Sugar refinery; 10 Flour milling; 11 Meat processing; 12 Dairy processing; 13 Other food; 14 Animal feed; 15 Large-scale farms: crop production; 16 Large-scale farms: Livestock production; 17: Small-scale farms: private subsidiary plots (LPH); 18 Small scale farms: Private farmers; 19 Trade & Transport; 20 Services.

Source: Updated version based of 1990 IOT (see Table A4-3).

References

Adelman, I. and S. Robinson (1978): *Income Distribution Policy in Developing Countries: A Case Study of Korea*. Oxford University Press, Oxford.

Banse, M. (1997): *Die Analyse der Transformation der ungarischen Volkswirtschaft*. Duncker & Humblot, Berlin.

Banse, M. and W. Münch (1998): 'Auswirkungen eines EU-Beitritts der Visegrad-Staaten: Eine partielle und allegemeine Geleichgewichtsanalyse'. *In*: Heißenhuber, A., H. Hoffmann, and W. von Urff (eds.): *Land- und Ernährungswirtschaft in einer erweiterten EU*. Schriften der GeWiSoLa e.V., Vol. 34, 301–309, Landwirtschaftsverlag, Münster-Hiltrup.

Banse, M. and S. Tangermann (1998): 'Agricultural implications of Hungary's accession to the EU – Partial versus general equilibrium effects'. In: Brockmeier, M., J.F. Francois, T.W. Hertel, and P.M. Schmitz (eds.): *Economic Transition and the Greening of Policies: Modeling New Challenges for Agriculture and Agribusiness in Europe*. Vauk Academic, Kiel: 1–21.

Berglöft, E. and R. Vaitilingam (eds. 1999): *Stuck in Transit: Rethinking Russian Economic Reform*. Centre for Economic Policy Research (CEPR), London.

Blanchard, O. (1998): *The Economics of Post-Communist Transition*. Clarendon Lectures in Economics. Clarendon Press, Oxford.

Braber, R. and F. van Tongeren (1996): 'Energy price reforms in Russia'. *Moct-Most*, 6: 139–62.

Braun, J. von and M. Qaim (1999): 'Household action in food acquisition and distribution under transformation stress'. In: Hartmann, M. and J. Wandel (eds.): *Food Processing and Distribution in Transition Economies: Problems and Perspectives*. Vauk Academic, Kiel.

Braun, J. von, H. de Haen, and J. Blanken (1991): *Commercialization of Agriculture under Population Pressure: Effects on Production, Consumption, and Nutrition in Rwanda*. IFPRI Research Report No. 85. Washington, D.C.

Braun, J. von, M. Qaim, and H. tho Seeth (2000): 'Poverty, subsistence production, and consumption of food in Russia: Policy implications'. In: Wehrheim, P., K. Frohberg, E. Serova, and J. von Braun (eds.): *Russia's Agro-food Sector: Towards Truly Functioning Markets*. Kluwer Academic, Boston/Dordrecht/London: 301–21.

Brooke, A., D. Kendrick, and A. Meeraus (1988): *GAMS – A User's Guide*. The Scientific Press, Redwood City.

Brooks, K. and Z. Lerman (1994): 'Land reform and farm restructuring in Russia'. Discussion Paper No. 233. The World Bank, Washington, D.C.

Bulmer-Thomas, V. (1982): *Input-Output Analysis in Developing Countries: Sources, Methods and Applications*. Wiley, New York/Chichester.

Business Information Service for the NIS (BISNIS; 2000): *Customs Update*. Moscow.

Central Bank of Russia (CBR; 1998): *Exchange Rate Policy*. Http://www.cbr.ru/eng/dp/pvk.htm, Moscow.

Chashnov, S. (1999): *Auswirkungen der Finanzkrise auf die Situation am Arbeitsmarkt*. Unpublished report for the research project 'Liberalization *cum* Decentralization in Russia'. Center for Development Research, University of Bonn, Bonn/Moscow.

Dervis, K., J. de Melo, and S. Robinson (1982): *General Equilibrium Models for Development Policy*. Cambridge University Press, Cambridge.

Devarajan, S., S. Go, J.D. Lewis, S. Robinson, and P. Sinko (1994): 'Simple general equilibrium modeling'. In: Francois, J.F. and K. A. Reinert (eds.): *Applied Methods for Trade Policy Analysis – A Handbook*. Cambridge University Press: 156–185.

Devarajan, S. and J.D. Lewis (1990): 'Policy lessons from trade-focused two-sector models'. *Journal of Policy Modeling*, 12: 625–57.

Devarajan, S., J.D. Lewis, and S. Robinson (1994): 'Getting the Model Right: The General Equilibrium Approach to Adjustment Policy'. Draft manuscript, January.

Dynnikova, O. (1999): 'Real appreciation and output – Russia 1993–1997'. Economics Education and Research Consortium (EERC), Working Paper Series, No. 99/13. Moscow.

Faini, R. (1988): *Elasticities of supply: some estimates for Morocco and Turkey*. The World Bank, mimeo, Washington, D.C.

Food and Agricultural Organization of the United Nations (FAO; 2000): *Trade Statistics*. FAO homepage (http//:www.FAO.org), Rome.

Fock, A., P. Weingarten, O. Wahl, and M. Prokopiev (2000): 'Russia's bilateral agricultural trade: First results of a partial equilibrium analysis'. In: Wehrheim, P., K. Frohberg, E. Serova, and J. von Braun (eds.): *Russia's Agro-food Sector: Towards Truly Functioning Markets*. Kluwer Academic, Boston/Dordrecht/London: 271–97.

Frohberg, K., M. Hartmann, P. Weingarten, O. Wahl, and A. Fock (1998): 'Development of CEEC agriculture under three scenarios: Current CEEC policies, CAP 1995/96, Agenda 2000'. In: Brockmeier, M., J.F. Francois, T.W. Hertel, and P.M. Schmitz (eds.): *Economic Transition and the Greening of Policies: Modeling New Challenges for Agriculture and Agribusiness in Europe*. Vauk Academic, Kiel: 236–56.

Gardner, B. and K. Brooks (1994): 'Food prices and market integration in Russia': 1992–93. *American Journal of Agricultural Economics*, 68 (5): 641–646.

Gaddy, C. and B. Ickes (1999): 'An accounting model of the virtual economy in Russia'. *Post-Soviet Geography and Economics* (formerly *Post-Soviet Geography*), 40 (2): 79–97.

Gatti, R. (1999): 'Corruption and trade tariffs, or a case for uniform tariffs'. *Macroeconomics and Growth*. Development Research Group, mimeo, World Bank, Washington, D.C.

Goodwin, B., T. Grennes and C. McCurdy (1999): 'Spatial price dynamics and integration in Russian food markets'. *Policy Reform*, 3: 157–193.

Goskomstat (Statistical Office of the Russian Federation) (1999): *Russia in Figures* (in English). Official publication, Moscow.

—(1998a): *Russian Agriculture* (in Russian: Selskoje chosjajstvo v Rossij). Official publication, Moscow.

—(1998b): *Statistical Yearbook* (in Russian: Rossiskij Statisticeskij Eshegodnik). Official publication, Moscow.

Government of the Russian Federation (2000): 'Basic Trends in Agrarian and Food Policy of the Russian Federation for 2001–2010'. Minutes No. 27, Session on June 27th, Moscow.

Haarparanta, P. and L. Kerkelä (2000): *Transition and Barter in Russia*. Mimeo, Department of Economics, Helsinki School of Economics.

Häger, G., D. Kirschke, and S. Nolepppa (2000): 'Empirische Bedeutung von Marktkräften im Transformationsprozess: Das Beispiel der Agrarproduktion in den Neuen Bundesländern'. *Agrarwirtschaft*, 49 (7): 252–59.

Hagedorn, K. (1998): 'Concepts of institutional change for understanding privatization and restructuring of agriculture in Central and Eastern European Countries'. In: Frohberg, K. and W.R. Poganietz (eds.): *The Importance of Institutions for the Transition in Central and Eastern Europe*. Vauk Academic, Kiel: 51–64.

Helmstädter, E., B. Meyer, E. Kleine, and J. Richtering (1983): 'Die Input-Output-Analyse als Instrument der Strukturforschung. Schriften zur angewandten Wirtschaftsforschung', 46, Mohr, Tübingen.

Hertel, T. (ed. 1997): *Global Trade Analysis – Modeling and Applications*. Cambridge University Press, Cambridge.

Henrichsmeyer, W. and Witzke H.P. (1991): *Agrarökonomische Grundlagen – Agrarpolitik*. Ulmer, Stuttgart.

Henrichsmeyer, W., O. Gans, and I. Evers (1991): Einführung in die Volkswirtschaftslehre, (9th edition). Ulmer, Stuttgart.

Ianbykh, R. (2000): 'Credit markets in rural Russia'. In: Wehrheim, P., K. Frohberg, E. Serova, and J. von Braun (eds.): *Russia's Agro-food Sector: Towards Truly Functioning Markets*. Kluwer Academic, Boston/Dordrecht/London: 409–27.

Institute of World Economics (IfW), Institute for Economic Research (IWH), and German Institute for Economic Research (DIW; 1998): 'Die wirtschaftliche Lage Russlands – Dreizehnter Bericht'. Kiel/Halle/Berlin. December.

International Monetary Fund (IMF; 1995): Russian Federation – Statistical Appendix. *IMF staff country report*, 107. Washington, D.C.

Ivanova, N. and G. Pavlov (1999): 'Gasoline Crisis in the Russian Federation: Russian Economic Trends'. Russian European Center for Economic Policy (RECEP), Monthly update, August.

Johansen, L. (1960): *A Multi-Sectoral Study of Economic Growth*. North-Holland, Amsterdam.

Johnson, S.R. (1998): 'Institutional versus policy reform in transition economies'. In: Frohberg, K. and W.R. Poganietz (eds.): *The Importance of Institutions for the Transition in Central and Eastern Europe*. Vauk Academic, Kiel: 17–30.

Jütting, J. and S. Noleppa (1999): *Agrarökonomische Forschung zu Strukturanpassung und Transformation: Mehr Gemeinsamkeiten als vermutet?* Mimeo, Bonn/Berlin.

Koester, U. (1998): 'Transforming socialist agriculture – from plan to market'. *European Review of Agricultural Economics*, 25 (3): 281–88.

Koester, U. and L. Striewe (1999): 'Huge potential, huge losses – The search for ways out of the dilemma of Ukrainian agriculture'. In: Siedenberg, A., L. Hoffmann (eds.): *Ukraine at the Crossroads: Economic Reforms in International Perspectives. Physica*, Heidelberg/New York.

Konings, J. and H. Lehmann (2000): *Going back to basics: Marshall and labor demand in Russia.* Draft manuscript presented at IZA Workshop on 'Microeconomics of labor adjustment in the Russian Federation', University of Bonn.

Krueger, A.O., M. Schiff, and A. Valdés (eds. 1991): *The Political Economy of Agricultural Pricing Policy* (Vol. 1). A World Bank Comparative Study. Johns Hopkins University Press, Baltimore/London.

Krylatykh, E.N. and I.V. Semyonova (1996): *The tax situation in Russian agriculture and agribusiness problems of improving land taxes.* Mimeo, Agrarian Institute, Moscow.

Kuhn, A. (2001): *Handelskosten und regionale (Des-)Integration: Russlands Agrarmärkte in der Transformation.* Peter Lang, Frankfurt/Main.

Kuhn, A. and P. Wehrheim (1999): 'Klassifizierung der russischen Agrar- und Ernährungswirtschaft in der Transformation auf der Ebene der Regionen'. *Europa Regional*, 7 (2): 2–9.

Langhammer, R. (2000): 'Die Europäische Gemeinschaft in der Entwicklung und in den Perspektiven des Welthandels zur Jahrhundertwende'. In: P.C. Müller-Graff (ed.): *Europäische Gemeinschaft in der Welthandelsorganisation.* Nomos, Baden-Baden: 19–46.

Lehmann, H., J. Wadsworth, and A. Acquisti (1999): 'Crime and punishment: Job insecurity and wage arrears in the Russian Federation'. *Journal of Comparative Economics*, 27 (4): 595–17.

Lewis, J.D., S. Robinson, and K. Thierfelder (1999): 'After the negotiations: Assessing the impact of free trade agreement'. Trade and Macroeconomics Division, Discussion Paper No. 46, IFPRI, Washington, D.C.

Lewis, W.A. (1954): *Economic development with unlimited supplies of labor.* Manchester School of Economic and Social Studies, 22 (2): 139–91.

Löfgren, H. (2000): 'Exercises in general equilibrium modeling using GAMS'. *Microcomputers in Policy Research,* No. 4 a, IFPRI, Washington, D.C.

Lohlein, D. and P. Wehrheim (2000): 'Rural development policies in Russia's transition process'. In: Buchenrieder, G. (ed.): *Timing and Sequencing of Reforms in Transition Economies.* Proceedings of a Mini-Symposium at the IAAE-Conference, Berlin.

Loy, J.P. and P. Wehrheim (1999): 'Spatial food market integration in Russia'. In: Peters, G. and J. von Braun (eds.): *Food Security, Diversification, and Resource Management: Refocusing the Role of Agriculture?* Proceedings of the 23rd International Conference of Agricultural Economists. Ashgate, Aldershot: 421–31.

Macours, K. and J.F.M. Swinnen (2000): 'A comparison of agrarian transition in Russia, China, and Eastern Europe'. In: Wehrheim, P., K. Frohberg, E. Serova, and J. von Braun (eds.): *Russia's Agro-food Sector: Towards Truly Functioning Markets.* Kluwer Academic, Boston/Dordrecht/London: 9–35.

Mansur, A. and J. Whalley (1984): 'Numerical specification of applied general equilibrium models: Estimation, calibration, and data'. In: Scarf, H.E. and J.B. Shoven (eds.): *Applied General Equilibrium Analysis*. Cambridge University Press, Cambridge.

Melo, J. de and S. Robinson (1989): 'Product differentiation and the treatment of foreign trade in computable general equilibrium models of small economies'. *Journal of International Economics*, 27: 47–67.

Melyukhina, O. and I. Khramova (2000): 'New players in the Russian food marketing chain'. In: Wehrheim, P., E. Serova, K. Frohberg, and J. von Braun (eds.): *Russia's Agro-food Sector: Towards Truly Functioning Markets*. Kluwer Academic, Boston/Dordrecht/London: 383–403.

Melyukhina, O., M. Qaim, and P. Wehrheim (1998): 'Regional protection rates for food commodities in Russia: Producer and consumer perspectives'. *European Review of Agricultural Economics*, 25 (3): 395–411.

Melyukhina, O. and P. Wehrheim (1996): 'Russian agricultural and food policies in the transition period: Federal and regional responsibilities in flux'. Discussion paper series *The Russian Food Economy in Transition*, No. 5, Institute for Food Economics and Consumption Studies, University of Kiel.

Michalopoulos, C. (1998): 'WTO accession for countries in transition'. Policy Research Working Paper No. 1934, The World Bank, Washington, D.C.

Mundlak, Y. (2000): *Agriculture and Economic Growth: Theory and Measurement*. Harvard University Press. Cambridge, Massachusetts/London.

Murphy, K., A. Shleifer, and R. Vishny (1993): 'Why is rent-seeking so costly to growth?' *American Economic Association, Papers and Proceedings*, 83 (2): 409–414.

North, D. (1995): 'The new institutional economics and third world development'. In: Harriss, J., J. Hunter, and C.M. Lewis (eds.): *The New Institutional Economics and Third World Development*. Routledge, London.

Nuppenau, E.A. and P. Wehrheim (1999): 'Benefits of dual market structures in the Russian meat market: Simulations and policy options'. In: Hartmann, M. and J. Wandel (eds.): *Food Processing and Distribution in Transition Economies: Problems and Perspectives*. Vauk Academic, Kiel: 131–48.

Obstfeld, M. and K. Rogoff (1996): *Foundations of Macroeconomics*. MIT, Cambridge, Massachusetts.

Organization for Economic Cooperation and Development (OECD; 2000a): *Agricultural Policies in Emerging and Transition Economies 2000*. Paris.

—(2000b): *Russian Federation, Economic Surveys*. Center for Cooperation with Non-Members, Paris.

—(1999a): *Russian Federation. Review of Agricultural Policies*. Center for Cooperation with Non-Members, Paris.

—(1999b): *Agricultural Policies in Emerging and Transition Economies 1999*. Vol. 1, Center for Cooperation with Non-Members, Paris.

—(1997): *Russian Federation. Economic Surveys*. Center for Cooperation with the European Economies in Transition, Paris.

—(1993): *National Accounts for the Former Soviet Union – Sources, Methods, and Estimates*. Center for Cooperation with the European Economies in Transition, Paris.

Pavel, F. (2000): 'Liberalization of a non-competitive market: Lessons from a forward-looking General Equilibrium Model for Bulgaria?' In: Heckelei, T., H.P. Witzke, and W. Henrichsmeyer (eds): *Agricultural Sector Modeling and Policy Information Systems*. Vauk Academic, Kiel.

Perrotta, L. (1999): 'Individual subsidiary holdings – The micro-economics of subsistence in Ukraine'. Center for Privatization and Economic Reform in Agriculture. Occasional Paper No. 12, Kyiv.

Poganietz, W.R. (2000): 'Inflation and exchange rate policy'. In: Wehrheim, P., K. Frohberg, E. Serova, and J. von Braun (eds.): *Russia's Agro-food Sector: Towards Truly Functioning Markets*. Kluwer Academic, Boston/Dordrecht/London: 129–153.

Pyatt, G. and J.E. Round (eds.) (1985): *Social Accounting Matrices: A Basis for Planning*. The World Bank, Washington, D.C.

Robinson, S. (1991): 'Macroeconomic, financial variables, and CGE models'. *World Development*, 19 (11): 1509–25.

—(1989): 'Multisectoral models'. In: Chenery, H. and T.N. Srinivasan (eds.): *Handbook of Development Economics*, Vol. 2, North-Holland, Amsterdam.

Rutherford, T. (1998): *Economic Equilibrium Modeling with GAMS – An Introduction to GAMS/MCP and GAMS/MPSGE*. GAMS Development Corporation, Washington, D.C.

Sadoulet, E. and A. de Janvry (1995): *Quantitative Development Policy Analysis*. John Hopkins University Press, Baltimore/London.

Salter, W. (1959): 'Internal and External Balance: The role of price and expenditure effects'. *Economic Record*, 35: 226–37.

Scandizzio, P.L. (2000): 'A computable general equilibrium model for a transition economy'. In: Poganietz, W.R, A. Zezza, K. Frohberg, and K.G. Stamoulis (eds.): *Perspectives on Agriculture in Transition: Analytical Issues, Modelling Approaches, and Case Study Results*. Vauk Academic, Kiel. 29–76.

Schmid, A.A. and P. Thompson (1999): 'Against mechanism: Methodology for an evolutionary economics'. *American Journal of Agricultural Economics*, 81 (5):1160–1165.

Sedik, D.J (2000): 'Discussant's comments on section 3'. In: Wehrheim, P., K. Frohberg, E. Serova, and J. von Braun (eds.): *Russia's Agro-food Sector: Towards Truly Functioning Markets*. Kluwer Academic, Boston/Dordrecht/London:180–183.

Sedik, D.J, M.A. Trueblood, and C. Arnade (2000): 'Agricultural enterprise restructuring in Russia, 1991–95: A technical efficiency analysis'. In: Wehrheim, P., K. Frohberg, E. Serova, and J. von Braun (eds.): *Russia's Agro-food Sector: Towards Truly Functioning Markets*. Kluwer Academic, Boston/Dordrecht/London: 495–512.

Seeth, H. tho (1997): *Russlands Haushalte im Transformationsprozess: Einkommens-, Armuts- und Versorgungsanalyse*. Series V, Economics and Management, Vol. 2098. Peter Lang, Frankfurt/Main.

Seeth, H. tho, S. Chachnov, A. Surinov, J. von Braun (1998): 'Russian poverty: Muddling through economic transition with garden plots'. *World Development*, 26 (9): 1611–23.

Serova, E. (2000): 'Russia's agro-food sector: State of the art'. In: Wehrheim, P., K. Frohberg, E. Serova, and J. von Braun (eds.): *Russia's Agro-food Sector: Towards Truly Functioning Markets.* Kluwer Academic, Boston/Dordrecht/London: 81–106.

Serova, E. (1998): *Impact of the economic crisis on the agricultural and food sector. Russian Economy: Trends and Perspectives.* http://www.online.ru/sp/iet/trends/sep98eng, Moscow.

Serova, E. and I. Khramova (2000a): 'Emerging supply chain management in Russia's agro-food sector'. Discussion Paper Series *The Russian Agro-food Economy in Transition,* No. 14, Center for Development Research, University of Bonn.

Serova, E. and I. Khramova (2000b): *Budgetary expenditures for agriculture in the Russian Federation in the 90s.* Institute for the Economy in Transition, mimeo, Moscow.

Serova, E., J. von Braun, and P. Wehrheim (1999): 'Impact of financial crisis on Russia's agro-food economy'. *European Review of Agricultural Economics,* 26 (3): 349–70.

Shleifer, A. and D. Treisman (1998): *The Economics and Politics of Transition to an Open Market Economy: Russia.* OECD Development Center Studies, Paris.

Shoven, J.B. and H. Whalley (1992): *Applying General Equilibrium. Cambridge Surveys of Economic Literature.* Cambridge University Press, Cambridge.

Steinberg, D. (1992): 'Economies of the Former Soviet Union – An input-output approach to the 1987 National Accounts'. *Working Papers, Socioeconomic Data,* World Bank, Washington, D.C.

Stoneman, R. (2000): *The prospects for economic growth and the consequences for poverty in Russia.* Mimeo, Russian European Centre for Economic Policy (RECEP), Moscow.

Swaminathan, P., Brockmeier, M., and T. Hertel (1998): 'Integration of Central and Eastern European Economies into the European Union'. In: Brockmeier, M., J. Francois, T. Hertel, and P.M. Schmitz (eds.): *Economic Transition and the Greening of Policies: Modeling New Challenges for Agriculture and Agribusiness in Europe.* Vauk Academic, Kiel.

Swan, T. (1960): 'Economic control in a dependent economy'. *Economic Record,* 36: 51–65.

TACIS-SIAFT (Technical Assistance to the Commonwealth of Independent States; SIAFT: *Support for Improving Agricultural Free Trade of the CIS;* various issues): Project Newsletter. http://www.aris.ru/ In both, English and Russian (Once at the home page follow the links for SIAFT). Moscow.

— (1999): Project Proposal. Mimeo, Bruxelles, Moscow.

TACIS (Technical Assistance to the Commonwealth of Independent States; 1998): *Farm structures in the Russian Federation.* Project Report, Moscow/Bruxelles.

Tarr, D. (1998): *The Design of Optimal Tariff Policy for Russia.* Mimeo, The World Bank, Washington, D.C.

Taylor, L. (1990). 'Structuralist CGE models'. In: L. Taylor (ed.): *Socially Relevant Policy Analysis. Structuralist Computable General Equilibrium Models for the Developing World.* Cambridge University Press, Cambridge.

Tesche, J. (1994): 'Alternative adjustments to external shocks in Hungary'. *Journal of Policy Modeling.*

Tillack P. and E. Schulze (2000): 'Decollectivization and restructuring of farms'. In: Wehrheim, P., K. Frohberg, E. Serova, and J. von Braun (eds.): *Russia's Agro-food Sector: Towards Truly Functioning Markets.* Kluwer Academic, Boston/Dordrecht/London: 447–70.

Troschke, M. and V. Vincentz (1995): 'Zuverlässigkeit und Problematik der statistischen Berichterstattung in Russland, Weissrussland und der Ukraine'. Working Paper No. 176, Osteuropa-Institut, München.

Varian, H.R. (1991): *Microeconomics*, (2nd edition). Norton Publishers, New York.

Walras, L. (1874 cf. in 1988): *Elements d'economie politique pure ou theorie de la richesse sociale.* Reprint with comments by von Hayek, F.A., M. Blaug, and W. Jaffe. Publisher Wirtschaft und Finanzen, Düsseldorf.

Wandel, J. (2000): 'Vertical Integration in the Russian Agro-food Sector'. In: Wehrheim, P., K. Frohberg., E. Serova, and J. von Braun (eds.): *Russia's Agro-food Sector: Towards Truly Functioning Markets.* Kluwer Academic, Boston/Dordrecht/London: 359–382.

Wehrheim, P. (2000a): 'The role of the agro-food sector in the macro-economy: General equilibrium effects'. In: Wehrheim, P., K. Frohberg, E. Serova, and J. von Braun (eds.): *Russia's Agro-food Sector: Towards Truly Functioning Markets.* Kluwer Academic, Boston/Dordrecht/London: 155–79.

— (2000b): 'Führt die Transformation zwingend zu mehr Armut?'. In: Diepenbrock, W., K. Frohberg, and J. Splike: *Land- und Ernährungswirtschaft in Mittel- und Osteuropa.* Tagungsband zum Welternährungstag 1999, Halle/Saale: 29–36.

— (1998): 'Institutional change in the Russian food marketing system'. In: Frohberg, K. and P. Weingarten (eds.): *The Significance of Politics and Institutions for the Design and Formation of Agricultural Policy.* Vauk Academic, Kiel: 176–91.

—(1997): 'Agrarian reform in Russia: Privatization and liberalization – The case of Pskov, Orel and Rostov oblasts'. In: Spoor, M. (ed.): *Privatization in Agriculture.* Intermediate Technology Publications, The Hague: 142–55.

Wehrheim, P. and J. von Braun (2000): 'Research implications: Beyond transition'. In: Wehrheim, P., K. Frohberg., E. Serova, and J. von Braun (eds.): *Russia's Agro-food Sector: Towards Truly Functioning Markets.* Kluwer Academic, Boston/Dordrecht/London: 523–534.

Wehrheim, P. and M. Wiebelt (1998): 'General equilibrium analysis of tax policy issues in Russia'. In: Brockmeier, M., J. F. Francois, T.W Hertel, and P.M Schmitz (eds.): *Economic Transition and the Greening of Policies: Modeling New Challenges for Agriculture and Agribusiness in Europe.* Vauk Academic, Kiel: 268–84.

—(1997a): 'Zur Konsolidierung des russischen Staatshaushalts: Eine allgemeine Gleichgewichtsanalyse finanzpolitischer Alternativen und deren Auswirkungen auf die Ernährungswirtschaft'. Discussion paper series *The Russian Food Economy in Transition*, No. 7, Institute for Food Economics and Consumption Studies, University of Kiel, Kiel.

—(1997b): 'Ökonomische Auswirkungen alternativer Steuersysteme auf russische Lebensmittelmärkte'. In: Bauer, S., R. Herrmann., and F. Kuhlmann (eds.):

Märkte der Agrar- und Ernährungswirtschaft. Schriften der Gesellschaft für Wirtschafts- und Sozialwissenschaften des Landbaues e.V., Vol. 33: 501–12.

Weiss, C. (2000): Book review on 'Transaktionskosten und institutionelle Wahl in der Landwirtschaft: Zwischen Markt, Hierachie und Kooperation' by V. Beckmann. *European Review of Agricultural Economics*, 27 (3): 406–408.

Weyerbrock, S. (1998): 'Reform of the EU's Common Agricultural Policy: How to reach GATT-Compatibility?' *European Economic Review,* 42: 375–411.

Wiebelt, M. (1996): 'Anpassung und Einkommensverteilung in Entwicklungsländern. Eine angewandte allgemeine Gleichgewichtsanalyse für Malaysia'. *Kieler Studien,* Horst Siebert (ed.), Vol. 276, Mohr, Tübingen.

Wiebelt, M., R. Herrmann, P. Schenk, and R. Thiele (1992): 'Discrimination Against Agriculture in Developing Countries?' *Kieler Studien*, Horst Siebert (ed.), Vol. 243, Mohr, Tübingen.

Wobst, P. (2000): 'Devaluation under decreasing marketing margins through infrastructure investment'. Trade and Macroeconomic Division, IFPRI, paper presented at the IAAE conference, Berlin.

—(1998): 'A 1992 social accounting matrix (SAM) for Tanzania'. Trade and Macroeconomics Division, Discussion Paper No. 30, IFPRI, Washington, D.C.

Wolf, S. (1996): *Begrenzter Erfolg der Lomé-Abkommen: Eine empirische Untersuchung der Wirkungen der EU-Zollpräferenzen auf den Handel der AKP-Staaten.* Series V, Economics and Management, Vol. 2002, Peter Lang, Frankfurt/Main.

World Bank (various issues): *World Development Indicators.* Washington, D.C.

—(1996): *From Plan to Market.* World Development Report 1996. Washington, D.C.

—(1995): *Input-Output-Tables for the CIS Republics.* Social Economic Division, mimeo, Washington, D.C.

—(1992): *Measuring the Incomes of Economies of the Former Soviet Union.* Policy research working series, WP No. 1057, Washington, D.C.

Zeddies, J. (2000): 'Organization of Russia's large scale farms'. In: Wehrheim, P., K. Frohberg, E. Serova, E., and J. von Braun (eds.): *Russia's Agro-food Sector: Towards Truly Functioning Markets.* Kluwer Academic, Boston/Dordrecht/London. 471–95.

Zentrale Markt- und Preisberichtstelle (ZMP; various issues): *Osteuropa – Agrarmärkte Aktuell.* Bi-weekly publication, various issues, Bonn/Berlin.

Author Index

Subject Index